BEGINNING STATISTICS

BEGINNING STATISTICS

Gene Zirkel
Nassau Community College

Robert Rosenfeld
Nassau Community College

McGraw-Hill Book Company

New York St. Louis San Francisco
Auckland Düsseldorf
Johannesburg Kuala Lumpur
London Mexico Montreal
New Delhi Panama Paris
São Paulo Singapore Sydney
Tokyo Toronto

BEGINNING STATISTICS

234567890 EBEB 43210

This book was set in Caledonia by Progressive Typographers. The editors were A. Anthony Arthur and Shelly Levine Langman; the designer was Jo Jones; the production supervisor was Charles Hess. The drawings were done by J & R Services, Inc.
Kingsport Press, Inc., was printer and binder.

Library of Congress Cataloging in Publication Data

Zirkel, Gene.
Beginning statistics.

Bibliography: p.
Includes index.
1. Statistics. I. Rosenfeld, Robert, joint author. II. Title.
HA29.Z48 519.5 75-6940
ISBN 0-07-072840-2

To
George,
David
and
Laura

CONTENTS

To the Student ix
To the Teacher xi

1 What This Book Is About—Intuitive Hypothesis Testing 1
2 Sampling and Random Sampling 12
3 Probability or Chance 24
4 Did the Expected Thing Happen? Chi Square 40
5 Testing Proportions—Binomial Experiments 75
6 Two-Sample Binomial Experiments 107
7 Averages and the Central-Limit Theorem 124
8 Comparing Averages 157
9 Estimation and Confidence Intervals 173
10 Correlation and Prediction 191
11 Distribution-Free Tests 209

Appendix A: Sample Data 243
Appendix B: Tables 247

B-1 Chi-Square Table 248
B-2 Square Roots 249
B-3a A Short Table of Areas Under a Normal Curve to the Left of z 257
B-3b A More Complete Table of Areas Under a Normal Curve to the Left of z 258
B-4 Critical Values of t for a Two-tailed Test 263
B-5 Critical Values of t for a One-tailed Test 263
B-6 Critical Values of the Coefficient of Correlation r_c for a Two-tailed Test 264
B-7 Critical Values of the Coefficient of Correlation r_c for a One-tailed Test 264
B-8 Critical Values for Rank Sum Test 265
B-9 Critical Values for Rank Correlation Coefficient r_s 265

Appendix C: Formulas 266
Appendix D: One-Tail Hypothesis Tests 272
Appendix E: Type II Errors 279
Appendix F: Bibliography 284
Appendix G: Selected Readings 286
Answers to Selected Assignments 294

Index 317

TO THE STUDENT

Statistical thinking will one day be as necessary for efficient citizenship as the ability to read and write.

H. G. Wells

This book is designed to help you learn some basic ideas about statistics. To accomplish this we are going to show you a *few* types of problems which are usually solved by applying statistical ideas. You will gain two specific skills: first, the ability to recognize situations which call for statistical reasoning and second, the ability to know what kinds of statistical methods are used in these situations. We hope the result will be that when you see statistics being used in affairs which affect your life, you will be in a better position to decide whether or not they are being used in a reasonable way.

One word of caution: this book is only an *introduction* to a few basic concepts. You will not be an expert when you get to the end of the book even if you understand it perfectly. We have included a list of more advanced books if you want to pursue the study of statistics further.

We have found that most students learn about statistics best when they are actively engaged in some kind of statistical experiment or project. So we are going to ask you frequently to answer questions, to join in class exercises and class discussions, to do experiments, to conduct surveys, and to collect random samples of various types. In short, this book is designed for a course in which the student does more than sit on a chair for the whole semester. We hope you enjoy this introduction to statistics.

Many places in this book have blanks to be filled in by you. If you intend to sell this book at the end of the semester, we suggest that you write on a separate sheet of paper and not directly in the book.

Acknowledgment
We are indebted to many people for their help with this manuscript. For critical comment we thank our colleagues at Nassau, especially Abe Weinstein, Doug Brown, George Miller, and Mike Steuer. For typing assistance we thank Debbie DeSanto, Kathy Lee, and especially Annette Stasinos for her interest and helpfulness. For inventiveness on the limericks we thank Sue Rosenfeld, Max Rosenfeld, and Pat Zirkel. For suggestions for character names we thank the management and guests of Maven Haven.

Gene Zirkel
Robert Rosenfeld

TO THE TEACHER

To know how to suggest is the great
art of teaching.
H. F. Amiel

This book is a combination of several ideas. First of all, we call it a work-text to stress the fact that it is primarily intended to be used in a classroom situation where time can be taken to perform little statistical experiments. It is not a workbook in the traditional sense of a large collection of problems to be practiced by the student. It is *not* especially suited to self-study. Nor is it especially suited to large lecture classes unless time is available in smaller groups to do the experiments and carry on the discussions.

Many teachers like to begin a lesson by giving a little information, and then asking the students leading questions to draw further information from their background or their intuition. Our class exercises make frequent use of this technique. Some are for the student to answer individually and others are to be used in class discussion led by you. These exercises also use class data in several instances. For example, how many students in this class own their own cars? In many cases the Teacher's Manual contains sample results that were actually obtained in one classroom.

It is helpful if an elementary text contains concise summaries of the basic ideas covered. This book does that. Where material has been introduced at length via an extended example, possibly including some class exercises, the ideas are finally collected and presented concisely in a step-by-step summary. Furthermore, each chapter ends with a glossary and a very brief chapter summary. Periodic review is also helpful for the beginning student. Therefore, starting with Chapter 5, there are review questions at the end of every assignment section.

Another idea we have made frequent use of is the capability of a computer to simulate a random sample from a given population. Instead of stating the mean "should be" near 30, or saying that *if* we did this many times we would get results such as 31, 34, 28, etc., we have simulated many experiments and printed the results. The teacher can avoid a lot of hand waving. The students can see what actually did happen when an experiment was repeated 100 times. They can see what kind of results usually occur and which results almost never happen. This provides a better grasp of the theories involved. We have found that this also increases the students' faith in the results of a hypothesis test—the highly improbable really does not happen very often, and the expected does usually occur. Students do not need to understand how to program a computer nor do they need access to such a machine; we have already run the programs and the results are printed in the book.

We have found that students who *do* statistics learn more than those who just read a textbook. Hence, assignments at the end of some chapters include experiments that students can design and carry out themselves using their own data, as well as the usual type of problems expected in a textbook. A supply of sample data is included in Appendix A. This was gathered from a random sample of students at Nassau Community College. It provides data for a host of original problems which can be designed by the instructor or the student as supplementary student projects.

This book is designed to get students involved in *doing* statistical experiments, rather than just reading about them. When we first assigned a student project, we learned that this helped the students get a better grasp of the material. As we taught the course again, we tended to get more student involvement, and each time the students seemed to understand the subject better. This book is the result of that approach. Additional information about student experiments and class exercises is found in the Teacher's Manual.

Since the experiments and projects that students can work on are necessarily simple, we have suggested in each chapter a set of readings relevant to the material in that chapter. We have used two main sources for the readings: *The New York Times*, because it is available in most college libraries, and *Statistics: A Guide to the Unknown*, edited by Judith M. Tanur, an excellent set of readings specially collected for courses like this one, under the joint auspices of the American Statistical Association and the National Council of Teachers of Mathematics. (The book is published by Holden-Day, Inc., 1972.) We strongly recommend that several copies be available to your students and that they read as many of the articles as possible. The readings will give a much broader idea of the scope of statistics than is likely to result otherwise.

We have also reproduced in Appendix G a few readings from less available sources. We hope that they can serve as a basis for class reports or discussions. It is our intention to show students that statistics really does have something to do with their lives; it has not been invented merely to keep their teachers employed.

We should also comment on the level and style of the book. The material is aimed at college students with relatively little mathematical background, although many in high school could use this book profitably. We make no use of algebra. Many of our students are neither highly prepared nor motivated for the course. We have deliberately written the book with a light hand and some of the examples are couched in the most absurd situations. We have taken the liberty to give the people and places in the book some outrageous names. There are two reasons for this: it makes the book more fun to read and, more importantly, it makes it easy to refer to the examples.

The book tends to be "wordy." This, too, is deliberate because we have found that most of the students who take the course can benefit from extra verbal explanation. We want the students to understand the concepts which lead to the formulas even though they are not presented with mathematical proofs or derivations of the formulas. The book is not rigorous at all. We feel there is nothing to be gained by rigor at this level.

The book aims at an *elementary* and intuitive understanding of statistics. For example, the difference between a one-tail test and a two-tail test is relegated to an appendix as is the treatment of Type II error. We hope that the students who finish this course can understand the broad outlines of the statistics which they meet in everyday life, rather than become familiar with the details that the expert statisticians handle.

We can make a few suggestions on using the book. We feel that any course should start off with the first five chapters. Chapter 6 depends on Chapter 5, and is the most popular topic for generating student projects. Chapter 8 depends on Chapters 6 and 7. If you do Chapter 9, but have skipped anything earlier, you will have to select only the appropriate materials out of Chapter 9. Chapters 10 and 11 are independent of each other and of Chapters 6 through 9.

We have prepared an extensive Teacher's Manual which contains further suggestions for presenting the material and conducting some of the class exercises. The manual also contains the answers to all the class exercises and to all the assignments.

Gene Zirkel

Robert Rosenfeld

ONE

What This Book Is About—Intuitive Hypothesis Testing

Let the chips fall where they may.
Anonymous

A STATISTICAL EXPERIMENT

Consider the following experiment. There is a large container. Into the container are put 500 red poker chips, 300 green poker chips, and 200 ivory poker chips. Except for color, all the chips are alike. The container is shaken so that the chips are thoroughly mixed. The purpose of the experiment is to find out what happens when 40 chips are picked at random. Dr. I. Pickem, the world famous statistician, reaches into the container and randomly picks (without looking) a sample of 40 chips. She records the results like this:

Red	Green	Ivory
23	13	4

Are these results amazing or surprising? Not really. You will notice that Dr. Pickem picked more red than any other color and more green than ivory. That is not surprising because there were more red chips in the container to begin with, and there were more green than ivory. Also notice that about half the sample was red (23 out of 40) which is not surprising since half the chips in the container were red.

Dr. Pickem puts the 40 chips back in the container and repeats the experiment for a second trial. The sequence of the results now looks like this:

Trial	Red	Green	Ivory
1	23	13	4
2	20	10	10

The results from trial to trial can vary a good bit. The results for each trial are unpredictable. Such an experiment is called a **random experiment.** By repeating the experiment, Dr. Pickem hopes to find out what *usually* happens. She wants to get some idea of how the results vary from one trial to the next. Therefore, she repeats the experiment many times to see which results are likely and which results are unlikely. After a full day of experimenting, Table 1-1 gives the list of her results for 50 trials.

TABLE 1-1

Trial	Red	Green	Ivory	Trial	Red	Green	Ivory
1	23	13	4	26	23	6	11
2	20	10	10	27	19	14	7
3	19	8	13	28	23	12	5
4	13	18	9	29	23	9	8
5	15	11	14	30	20	10	10
6	21	12	7	31	23	4	13
7	17	12	11	32	17	17	6
8	21	11	8	33	21	11	8
9	20	11	9	34	19	16	5
10	26	8	6	35	27	7	6
11	11	20	9	36	15	14	11
12	23	7	10	37	24	11	5
13	23	12	5	38	20	15	5
14	26	10	4	39	22	6	12
15	20	12	8	40	20	18	2
16	17	12	11	41	19	14	7
17	18	13	9	42	23	9	8
18	23	9	8	43	21	8	11
19	24	9	7	44	14	18	8
20	15	16	9	45	26	11	3
21	24	10	6	46	23	7	10
22	20	13	7	47	23	11	6
23	22	12	6	48	24	9	7
24	20	10	10	49	23	12	5
25	26	11	3	50	23	5	12

CLASS EXERCISE*

Here are some outcomes which did not occur. Are they impossible? Why do you think they did not occur?

1-1 40 red 0 green 0 ivory ________________

1-2 0 red 20 green 20 ivory ________________

1-3 0 red 0 green 40 ivory ________________

1-4 5 red 10 green 25 ivory ________________

1-5 20 red 9 green 11 ivory ________________

1-6 12 red 13 green 15 ivory ________________

The next day Dr. Pickem is given the same container and is asked to decide quickly

* It is suggested that these exercises be discussed by the class at this point. Throughout this book questions will often be posed and they should be answered by the readers *before* they read on.

if the container still holds 500 red, 300 green, and 200 ivory chips. She immediately takes a random sample of 40 chips and gets

Red	Green	Ivory
12	13	15

1-7 What do you think she decided? ______________________

1-8 Can she be sure that she is right? ______________________

Statistical methods are often used to help an investigator decide whether a numerical statement about a population is probably true or probably false. In statistics a **population** is any group of things (possibly people) that the statistician is interested in. Usually, however, the investigator does not have access to the entire population, but only part of it, which we call the **sample.** When the investigator has some sample data but not complete data about the entire population, the statement about the population is called a **statistical hypothesis.**

In Dr. Pickem's experiment the hypothesis is: The container holds 200 ivory, 300 green, and 500 red chips. Dr. Pickem does not know *for sure* whether this hypothesis is true or false, and since she cannot examine the entire contents of the container, she *will not be able* to decide with perfect confidence whether the hypothesis *is* true or false. In fact, in most statistical experiments, the investigator *cannot* be perfectly sure whether the hypothesis is true or false. The job is rather to decide whether it is *probably* true or false and then to justify that decision.

In the example above, Dr. Pickem decided that the contents of the container had been changed, and she is probably right. Why? Because based on what she knew about the container from her previous experiment, she knew that the result 12 red, 13 green, and 15 ivory was *very* unlikely for a container with 500 red, 300 green, and 200 ivory chips.

Consider this analogous example. In the town of Grafton (where Dr. Pickem lives, of course), there are 1,000 registered voters. Of these, 500 are Republicans, 300 are Democrats, and 200 are Independents. A committee, called the Town Committee, is supposed to be formed by picking the names of 40 citizens at *random* from the list of registered voters.

The new membership of the committee is announced, and it is found to contain 30 Republicans, 5 Democrats, and 5 Independents.

CLASS EXERCISE

1-9 Do you think the committee *was* picked randomly? ____________

1-10 Can you be sure? ____________

Dr. Pickem would decide that the committee has *not* been picked randomly. Why? Because this problem is just like the experiment with the chips, and such a result is *very unlikely* to occur when the selection *is* random.

EXPERIMENT TWO

Your campus newspaper claims that 53 percent of students on this campus own cars. This claim can be considered as a statistical hypothesis. Suppose you wanted to determine if it is probably true or probably false. If *you* gathered a *random* sample of 200 students and found that only 98 owned cars, it would mean that only 98/200 or 49 percent of your sample owned cars.

CLASS EXERCISE

1-11 Is it *possible* that 53 percent of the entire campus population own cars and yet only 49 percent of the 200 students that you sampled own cars? ____________

1-12 Would you think the newspaper's hypothesis was probably right? ____________

1-13 If you had interviewed 500 students instead of 200, would that change your confidence in your answer to Exercise 1-11? ____________

1-14 If you had interviewed 50 instead of 200, would that change your confidence in your answer to Exercise 1-11? ____________

1-15 Suppose you had sampled *all* of the students? ____________

1-16 Suppose your sample of 200 students had only 25 percent car owners. What conclusions would you draw about the original claim that 53 percent of students own cars? ____________

1-17 Compute the percentage of the students in this room who own their own cars.

Number of students in the room ____________

Number who own a car ____________

Percentage who own a car ____________

1-18 Does this information tend to support or deny the claim that 53 percent of the students at this school own cars? ______________________________

EXPERIMENT THREE

Suppose that the *average income* for workers who have full-time jobs in the town of Wilslip is *known* by the IRS to be $10,730. A college student gathers a sample of workers at his favorite bar and discovers that the average salary in *his* sample of 26 full-time workers is $11,528. His teacher (after asking to see the list of 26 salaries) claims that the student's sample was *not* a *random* sample of the workers in the town of Wilslip.

CLASS EXERCISE

1-19 *Could* this sample be *random*? ______________________________

1-20 Is it very likely that it is random? Give some reasons. ______________________________

____

1-21 Make a list containing the amount of money each student in this class earned by working last year. ______________________________

1-22 Compute the *average* amount of money earned last year by students in this class. ______________________________

1-23 Do you think that the answer to Exercise 1-22 is representative of the average amount of money earned last year by *all* the people who are now students at your campus? Give reasons. ______________________________

EXPERIMENT FOUR

If a *fair* coin is tossed 7 times, it should come up heads about 3 or 4 times. In fact, it can be calculated that on the basis of statistical theory:

No heads will appear about 1 percent of the time
1 head will appear about 5 percent of the time
2 heads will appear about 16 percent of the time
3 heads will appear about 27 percent of the time
4 heads will appear about 27 percent of the time
5 heads will appear about 16 percent of the time
6 heads will appear about 5 percent of the time
7 heads will appear about 1 percent of the time

CLASS EXERCISE

Have each person in the class toss a coin 7 times and count the number of heads he gets.

1-24 Let S equal the number of times *you* were successful in getting heads. Record your value of S. ______________________

1-25 Let n equal the number of persons who tossed a coin. $n =$ ____________

1-26 Record class results in Table 1-2. Count how many persons got $S = 0$ and enter this number in column B. Then ask how many persons got $S = 1$, etc., until you have entered eight numbers in column B.

TABLE 1-2

A Success Equals Number of Heads	*B* Frequency Equals Number of Persons	*C* Percent of Persons	*D* Theoretical Percent
0			1
1			5
2			16
3			27
4			27
5			16
6			5
7			1

1-27 To get the percentage in column C, divide each number in column B by n. We expect about 54 percent of the class to toss 3 or 4 heads.

1-28 What percentage did toss 3 or 4 heads? ______________________

1-29 If it is not 54 percent, how can you explain the difference? ____________

1-30 Sue tosses a coin 7 times and gets 7 heads. She then claims that the coin was not a fair coin but was biased for heads. Would you say that she is:

a Certainly right
b Probably right
c No more likely to be right than wrong
d Probably wrong
e Certainly wrong

COMPUTER SIMULATION

At many places in this book we will talk about a statistical experiment and mention what might happen if this experiment were repeated many times. For example, you, along with each member of your class, have just tossed a coin 7 times. Probably, your class contains about 20 to 30 students. You might wonder what the results would be if there were, say, 500 or 3,000 people each tossing a coin 7 times. It is easy to use an electronic computer to simulate (or imi-

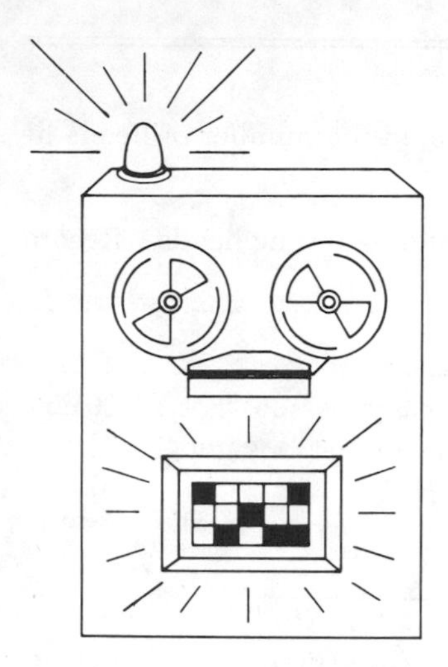

tate) this situation. It is possible to program a computer so that the numbers it calculates can be interpreted as tosses of a coin. For example, a computer can scramble numbers into an apparently random order, and then we can interpret them as follows.

If a number is *even,* it represents "heads," while *odd* represents "tails." It is very easy and quick to get a list of 7 "coin tosses" this way. And it is very easy to repeat this experiment a large number of times. Printed in Table 1-3 is the final result of a *simulated* coin toss where each of 500 "people" tossed a "coin" 7 times. The 500 results are printed in the order they occurred. The first three numbers in the printout are 2, 0, 3. This means that the first person

TABLE 1-3 COMPUTER SIMULATION OF 500 PEOPLE EACH TOSSING A COIN 7 TIMES AND RECORDING THE NUMBER OF HEADS TOSSED

2	0	3	4	7	2	5	4	3	4	7	5	3	4	2	6	5	3	5	3	4	7	3
2	3	5	5	6	4	4	6	2	6	5	5	4	2	2	4	3	2	3	2	4	3	3
4	5	5	3	4	2	2	3	1	3	4	3	3	3	2	4	3	4	4	4	3	1	2
4	4	4	1	3	4	4	3	2	5	4	3	5	4	5	3	3	4	6	2	6	4	2
5	4	4	5	3	4	5	4	7	2	5	3	2	3	4	5	1	4	6	3	2	4	4
3	4	5	3	4	5	5	3	6	4	3	4	1	3	4	5	4	5	2	1	3	4	4
6	5	6	3	4	5	3	3	5	3	4	4	5	5	5	1	4	4	3	3	1	5	2
4	2	3	2	5	4	5	5	5	3	4	3	3	3	3	5	3	1	2	6	3	5	4
2	2	3	3	3	4	5	3	5	4	6	3	4	5	2	5	3	2	3	2	2	4	2
5	6	2	3	3	3	3	5	7	5	4	6	4	4	3	4	3	4	3	3	3	4	2
4	0	5	3	3	6	5	3	3	4	2	6	3	5	5	5	5	2	3	2	4	4	4
3	4	1	4	3	3	3	4	6	1	3	3	4	3	4	4	5	2	5	4	2	3	3
4	5	5	3	2	3	4	4	2	3	5	3	2	4	4	5	5	4	4	3	5	6	3
4	4	4	3	1	3	4	4	3	3	6	2	3	5	3	1	5	5	4	2	4	3	5
1	4	4	4	4	4	5	4	4	4	5	5	5	2	2	3	4	3	2	5	4	3	6
2	1	1	5	7	3	2	4	4	4	6	5	4	1	5	3	4	4	1	3	4	4	3
3	5	4	4	3	2	5	5	2	2	3	3	3	5	5	5	4	2	5	5	3	6	3
5	1	2	4	1	6	1	5	2	4	4	6	3	3	3	2	4	3	2	4	5	5	3
2	5	2	1	3	1	2	5	3	1	6	4	4	2	4	2	5	6	3	1	4	2	4
4	2	2	3	6	2	5	5	3	2	6	3	3	3	3	4	3	4	3	2	3	2	5
3	3	3	3	3	1	5	2	3	0	4	4	6	3	2	2	3	3	3	3	2	3	3
5	5	5	4	1	3	4	5	3	4	3	5	4	3	4	2	4						

had 2 heads and 5 tails. The second person had *no* heads and 7 tails. The third person had 3 heads and 4 tails.

You can see that the results for any one person are unpredictable. There is no apparent pattern to the results. This is what you would expect to happen in a real experiment. These results, however, are difficult to interpret, precisely because they are not organized. In Table 1-4, therefore, we have grouped the various outcomes and counted them. The frequency of each outcome is printed for easier reading. For example, the second line of the chart says that 27 people out of 500 tossed 1 head and 6 tails, and that 27/500 is approximately 5 percent of the people in this experiment.

TABLE 1-4

S	Frequency	Approximate Percent	Approximate Percent Predicted by Theory
0	3	1	1
1	27	5	5
2	73	15	16
3	140	28	27
4	128	26	27
5	94	19	16
6	29	6	5
7	6	1	1

If you examine the results of this computer simulation, you will see that the results are very close to those predicted by mathematical theory. For example, the predicted percentage of people tossing 3 heads was 27 percent, and the simulation produced 28 percent. So this computer simulation was useful in two ways. First, because the results agree with theory, it increases our confidence in the theory. Second, it actually shows us a typical string of 500 outcomes in a way which is much more convenient than having someone actually toss a coin 7 times and repeat this 500 times.

Here is an example which illustrates another valuable use for computer simulation. Suppose you are asked to play the following game. Take 10 cards numbered from 1 to 10. Shuffle them. Hold them face down. Turn over the first one and say "one." Turn over the second one and say "two." Continue until you get to the tenth card. You win money as follows: Whenever the number that you say is actually the number that shows on the card, you win that many dollars. For example, if you turn over a "two" on the second card, and a "seven" on the seventh card, then you win $2 + 7 = 9$ dollars.

The question is this. How much money can you expect to win if you play this game often? You might not know enough mathematical theory to know what to predict. But this game can easily be simulated. All we need to do is have the computer print out a sequence of 10 numbers from 1 to 10 in random order over and over again. We would be able to see how much money we would usually win at this game.

In conclusion, an important use of simulation is to help an experimenter *figure out some theory* by providing a large list of typical outcomes for him to analyze.

You might be interested to know what actually happened when we simulated the card game 1,000 times. We found that on the average we won $5.29. You might like to play the game several times yourself to see how well you do. Can you figure out any theory for how much you expect to win?

Chapter Summary

We have examined briefly three kinds of statistical situations.

1 The political committee (or the poker chip) problem is an ex-

ample of a **chi-square experiment.** It asked, "Did the *difference* between the *expected* numbers of persons from each party on the committee and the actual *observed* numbers occur randomly or was there a systematic exclusion of some people?"

2 The car ownership problem is an example of a **binomial experiment.** It asked whether the *proportion* of car owners in a random *sample* was significantly *different* from the proportion that was supposed to be correct for the entire *population.*

3 The workers' income problem is an example of a **sample-mean experiment.** (Here the word "mean" indicates *average.*) The experiment asked whether or not a given sample could be considered random on the basis of the *difference* between the average of the *sample* and the average of the *population.*

We will study these and other related problems in this book. We shall learn how to recognize each of these types of experiment, how to carry out each type of experiment, and how to interpret the results of each type of experiment.

Words You Should Know

Computer simulation
Population
Random experiment
Sample
Statistical hypothesis

ASSIGNMENT

A salesman, one of the sliest
"Proved" that his brand rated highest
He collected a sample
And the size was most ample
But his method no doubt was quite biased.

1 "Warning, the Surgeon General has determined that cigarette smoking is dangerous to your health." Discuss this warning. How can the Surgeon General arrive at this conclusion?

2 Bring to class an example of statistics in a newspaper, magazine, or textbook.

3 Complete this sentence: The larger a random sample is, the (more) (less) accurately it will tend to represent the population.

4 Read Chapter 1 in Darrell Huff's droll little book, "How to Lie With Statistics" (Norton, New York, 1954), in order to learn the difference between a biased sample and a random sample.

Readings from J. Tanur (ed.), "Statistics: A Guide to the Unknown," Holden-Day, San Francisco, 1972.

5 Statistics, Scientific Method, and Smoking, Brown, p. 40. What are some of the statistical arguments for and against identifying smoking as a cause of cancer? What is a "control group"? What is the difference between a retrospective and a prospective study?

6 Deathday and Birthday: An unexpected connection, Phillips, p. 52. Can some people postpone their own deaths? What percent of the people who are born in September should die in October? Does it actually work out that way?

7 How Accountants Save Money by Sampling, Neter, p. 203. Can you really get an accurate description of a population just by looking at a sample? Is it worth taking a chance?

Readings from The New York Times

8 Read about the difference between a relationship and a cause in "Smoking is linked to loss of time from work and recreation," May 2, 1967, p. 16.

9 Examine the advertisement "The Most Exciting Headache News in Years!" September 20, 1970, section 2 (arts and leisure), p. 27.

10 Read about an argument over *poor* sampling in the New York City schools on p. 25B, May 26, 1971.

TWO

Sampling and Random Sampling

The preceding recorded message was selected from random phone calls.

A radio commercial for soup

You trust your mother, but you cut the cards anyway. Why do we shuffle cards before we deal? We want them to come out *randomly*. We want each person to have the same chance of winning whatever game is being played. The idea of randomness is very important in statistics. We say that a card is selected at random from a deck if every single card has an equal chance of being picked. We say that a raffle ticket is picked at random if every single ticket that was purchased has the same possibility of being selected. If Dr. Pickem picks a sample of 200 people out of a population of 9,000 persons, we say that her sample is a random sample if each one of the 9,000 people had the same chance to be in the sample.

Very often it is impossible for an experimenter to obtain a perfect random sample. People who gather opinion polls must often be satisfied with a less than perfect random selection method. But if the selection process used is carefully designed to be free of bias, it can often be considered random, and much useful information may be gathered from such a sample.

CLASS EXERCISE

A team of sociologists collected the following data on alcoholism:

Number of people interviewed: 100
Number who said they were alcoholics: 22

What conclusions can we draw from these statistics?

2-1 Write down one conclusion you would make on the basis of this data. I would conclude that ____________________________

Here is a list of some conclusions that other students have suggested:

A 22 percent of Americans *are* alcoholics.
B 22 percent of Americans *admit* to being alcoholics.
C Probably less than half the people are alcoholics.
D About 22 percent of the men in the United States are alcoholics.

2-2 Would you agree with any of these conclusions? ____________________

Why? ____________________________________

Here is a much longer list of objections some students have raised when they were asked to answer the question.

A This is ridiculous. You can't tell anything from just 100 people.
B Where did they get these people?

C Just because 22 *said* they were alcoholics doesn't mean much, maybe some were afraid to admit it.

D Were these people *all* men, or what?

E Were they all college students? Lots of experiments are only tried on college students.

F *Who* asked the people. I mean did men interview men, or did men interview women, or what?

G What do you mean by alcoholic, anyway? Suppose a person drinks a lot but it doesn't bother him?

H Were the people running the interviews alcoholics themselves?

I Why did they do this experiment in the first place?

Perhaps, you raised one or more of these objections yourself.

2-3 Can you think of any other objections? ______________________________

____

All of these objections are valid, and anyone who does a statistical experiment has to keep problems like these in mind. We hope that by using this book you will become very critical of statistical experiments and learn to distinguish between good ones and bad ones.

When someone looks at data based on some number of things or people and then announces a conclusion about a larger number of people or things, we say that he has made an **inference.** For example, suppose you look at 50 cars on your campus and notice that 25 are foreign cars. If you decide that probably half of *all* cars on *all* campuses are foreign cars, then you have made an *inference.* A major portion of this textbook will be concerned with making good inferences. The group of people or items which you observe, the ones on which you base your inference, is called the **sample.** The larger group which you think the sample represents is called the **population.** For you to be able to make a "good" inference, you will need to choose a "good" sample.

Perhaps the two most important criteria for picking a "good" sample are: (1) It must be from the correct population. You must design your experiment so that you know what the population is before you begin to collect data. (2) The sample should be picked randomly, not in any biased way.

CLASS EXERCISE

2-4 Consider that your population is *all* the students in this class. Suggest a way that you could pick three of them randomly. This would *be a random sample of* *size 3* of the students in this class. ______

2-5 Is the method you suggested in answer to Exercise 2-4 a practical method for obtaining a random sample of size 100 from the entire body of your school?

______

2-6 Is this class a random sample of the student population of this school?

______

2-7 What role does the size of the population play when we are looking for a practical way to gather a random sample? ______

In Exercise 2-6 there is no answer, of course, until we know the purpose of the sample. A class may be a very good random sample of the school for the purpose of estimating shoe size, but may not be random at all if we are interested in estimating the percentage of biology majors on campus.

Note also that a sample does not have to be large in order to be random. A sample of two people may be random, for example, the winners of two prizes in a raffle. Sample size, of course, does play a role when we try to make inferences about a population based solely on data from a random sample of that population. We would be more confident in our inferences if they were based on data from a *large* sample.

CLASS EXERCISE

Select the first *five* people on the alphabetic roster of this class. These people are a sample of this class.

2-8 Are they a *random* sample? ______

2-9 If we compute the average height of these five people, will that be a good estimate for the average height of the entire class? ______

2-10 Compute this sample average. ______

2-11 Use it to estimate the class average. ______

2-12 Then compute the class average. ______

2-13 Is the sample average close to the class average? ______

2-14 If two of these first five people are the same nationality, for instance Italian, could we say that about $\frac{2}{5}$ of the class is Italian? (That is, are the first five names a good random sample of ethnic background?) ______

2-15 Consider the following questions:

A What percentage of the students on this campus are chemistry majors?
B What is the average age of teachers on this campus?
C Which course do students find easier, first semester history or first semester English?

Pick one of these questions and discuss how you would gather a random sample of 100 people at your school to estimate an answer to the question. Be *very specific* and also be practical. Where could you go to get an unbiased sample? Should you consider part-time students? What *exactly* would you do? Who would you ask? What *specific* questions would you ask? How would you record the data? What are some of the good points of your procedures? What are some of its weaknesses? Here is an example:

Question: Outline a method of estimating the percentage of the employees at the local Grimms Brothers factory who support the village ordinance outlawing beards and mustaches on policemen.

Answer: I will stand outside the main employees entrance at 5 P.M. on Thursday evening and interview every fifth worker leaving. If a group of employees are together, I will interview only the one on the left. If someone refuses to answer, I will ask the next person. I will continue until I get 50 responses. I will say:

1 Pardon me, I am doing a survey for my college class in statistics, would you answer two questions?
2 Are you employed here?
3 Do you favor the ban on policemen wearing beards and mustaches?
4 Thank you.

I will record the responses as follows:

Person	Yes	No	Undecided	Refused to Answer	Not in the Population
1					
2					
3					
. . .					
50					

Since I am avoiding a Monday or a Friday, fewer employees will be absent. Picking only one in a group avoids the bias of peer pressure. I do not know why employees leave by one exit rather than another; this might bias my sampling.

2-16 A fair die is supposed to produce a number from 1 to 6 randomly. Suppose you had a die that was *not* fair, but was loaded or weighted slightly in favor of the number 6. Describe a procedure for detecting this bias. Describe how you would collect data and use the data to try to convince an impartial judge that you were correct. ______________________________

____

COMPUTER SIMULATION

We wrote one computer program to simulate a *fair* die, and another program to simulate a die *biased* slightly in favor of 6.

A fair die tossed 60 times would be expected to produce about 10 of each possible result: The biased die was designed to produce about 9 ones, 9 twos, 9 threes, 10 fours, 10 fives, and 13 sixes in 60 tosses.

To put this experiment in the framework of sampling statistics, we first need to define the populations. The first population is *all* the outcomes we would get with the fair die if we would toss it forever. The second population is all outcomes for the biased die if we tossed it forever. These are clearly both infinite populations, and we must be satisfied with only samples from them.

For our first sample (the fair die), we ran our computer program to simulate 60 tosses, and we got the printout in Table 2-1 for our outcome.

TABLE 2-1 60 TOSSES OF FAIR DIE

```
5 3 5 4 1 1 3 3 3 5 1 3 4 2 3 6 3 2 6 4 4 4 6
6 5 4 3 3 3 3 2 1 6 5 4 4 2 6 5 3 1 4 2 5 6 2
5 6 3 3 4 5 5 1 3 2 1 5 2 6
```

Summarizing the above in a frequency chart, we have

Number Showing on Die	1	2	3	4	5	6
Frequency	7	8	15	10	11	9

Summarizing with a frequency graph will look like Figure 2-1. In theory we expected a graph like the one in Figure 2-2.

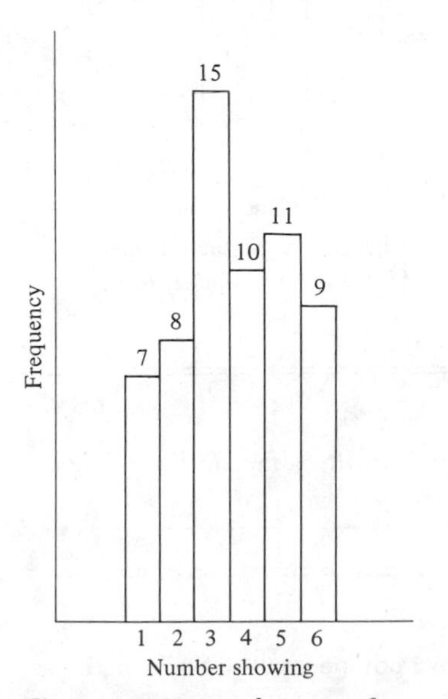

Figure 2-1 60 actual tosses of a fair die.

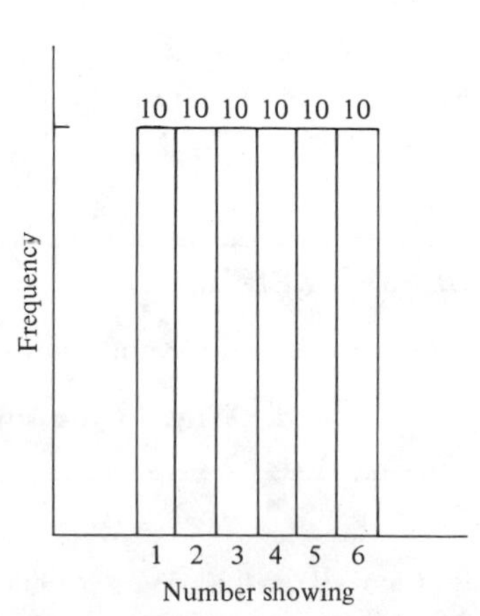

Figure 2-2 Expected results for 60 tosses of a fair die.

For the second sample (the biased die), we got the outcome shown in Table 2-2.

TABLE 2-2 60 TOSSES OF BIASED DIE

5	4	6	4	5	4	4	2	2	4	1	6	5	6	3	1	2	6	2	6	2	3	3
3	1	2	4	2	2	4	1	3	4	1	2	5	6	1	5	1	3	3	1	2	1	4
2	4	4	5	3	5	6	6	6	1	5	2	5	1	2								

Summarizing in a frequency chart, we have

Number Showing on Die	1	2	3	4	5	6
Frequency	11	13	7	11	9	9

Summarizing in a frequency graph, we will have the results shown in Figure 2-3. In theory we expected a graph like the one in Figure 2-4.

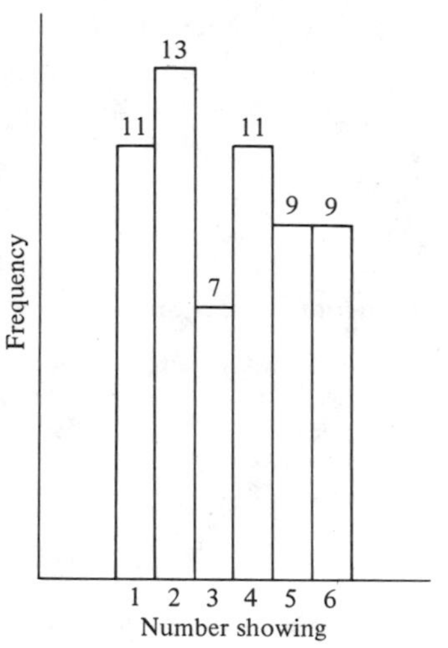

Figure 2-3 60 actual tosses of a biased die.

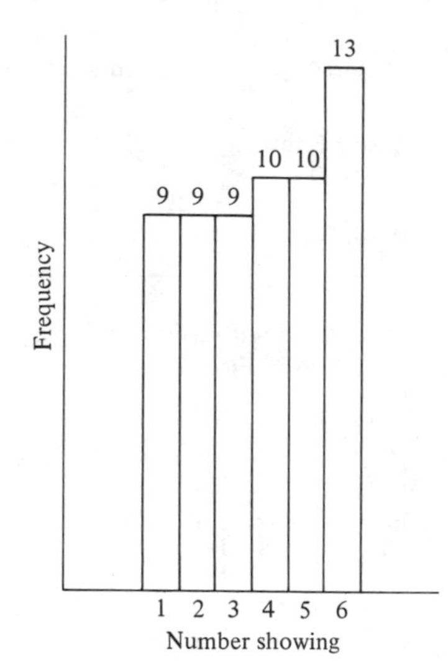

Figure 2-4 Expected results for 60 tosses of a biased die.

CLASS EXERCISE

2-17 It is certainly not obvious from this data that the first die is fair and the second biased. Why do you suppose this happened? ____________________

Then we ran the program again to simulate 600 tosses. We got the results shown in Tables 2-3 and 2-4.

TABLE 2-3 600 TOSSES OF FAIR DIE

2 2 3 4 4 5 4 5 6 2 2 3 4 5 1 3 5 6 3 6 3 6 6 4 3 3 5 1 2
5 1 2 1 3 3 4 6 3 4 5 2 5 2 4 5 3 6 6 3 4 4 4 6 4 3 6 4 1
5 4 5 4 4 6 6 3 2 5 4 6 2 5 6 2 6 2 6 4 5 1 1 4 6 6 5 2 6
3 4 2 3 1 2 6 5 4 3 2 5 6 3 6 3 2 4 5 3 4 4 2 4 2 2 4 2 5
5 6 4 4 2 5 3 3 1 3 6 4 6 5 3 5 2 5 1 4 1 1 5 6 5 2 3 6 6
6 1 3 2 2 6 5 1 3 3 1 3 3 1 1 6 5 1 4 3 6 2 4 2 1 6 6 1 2
5 1 6 6 3 2 2 5 4 2 1 1 5 3 6 5 6 4 1 3 3 3 5 3 1 4 2 5 4
5 6 3 3 2 3 3 1 1 6 5 3 4 2 6 3 3 1 2 1 5 6 5 4 5 2 6 2 5
6 1 5 4 2 5 1 3 3 3 5 5 1 4 1 1 4 2 6 2 1 3 1 6 3 2 3 6 5
4 6 5 1 2 3 5 4 1 2 5 5 4 6 3 6 5 4 1 3 5 1 1 2 3 1 4 6 3
2 5 3 2 6 5 4 4 5 2 5 3 3 5 3 2 5 4 6 5 3 6 1 3 6 5 5 5 5
5 4 5 5 4 4 2 4 1 3 6 2 4 4 4 2 1 4 5 6 4 1 1 6 1 6 6 3 2
5 5 5 2 4 4 1 1 2 3 3 5 6 4 4 6 2 1 2 5 6 4 2 1 6 6 1 6 3
3 1 6 3 6 3 6 6 6 3 3 5 3 2 3 2 3 3 5 6 3 5 6 3 6 3 6 6 4
3 2 6 6 1 4 6 3 2 3 5 4 5 3 2 6 2 2 1 2 1 5 5 4 1 5 1 5 5
4 4 5 4 5 3 2 4 1 1 4 3 4 2 1 3 4 5 3 6 6 1 2 4 1 3 2 4 1
5 6 5 4 6 3 5 2 3 1 1 1 2 1 3 4 3 2 4 6 4 4 1 2 2 4 5 1 3
3 2 2 1 2 2 5 4 4 4 6 1 3 1 6 6 5 5 1 3 5 4 1 1 3 5 3 3 2
3 1 5 2 1 3 6 4 2 1 1 4 4 6 2 5 2 4 5 5 4 5 4 1 2 4 2 4 6
1 5 1 4 3 4 5 2 2 4 3 5 5 3 5 1 3 6 5 4 1 5 2 5 1 4 6 6 5
6 5 6 3 1 2 6 3 1 1 3 4 2 1 5 5 3 3 2 1

Summarizing Table 2-3 by a frequency chart, gives

Number Showing on Die	1	2	3	4	5	6
Frequency	92	92	109	100	110	97

TABLE 2-4 600 TOSSES OF BIASED DIE

3 4 1 5 3 3 5 1 6 3 3 3 1 2 4 2 4 4 6 3 5 1 1 6 2 6 3 4 3
2 1 4 6 6 2 4 2 6 5 4 6 1 3 5 6 6 6 5 1 2 5 4 6 2 2 4 2 6
5 4 5 6 6 5 3 4 4 4 2 4 2 6 6 1 5 1 3 2 4 6 3 6 4 2 4 4 3
5 3 5 5 6 5 5 3 2 1 3 2 5 6 6 4 5 3 1 2 5 1 4 6 1 2 5 5 2
1 5 3 4 2 3 6 5 6 5 4 6 3 5 2 6 2 2 2 3 6 4 1 4 1 5 4 3 5
5 1 4 1 3 6 3 6 4 4 4 2 3 6 2 5 6 3 4 1 6 5 2 1 2 1 3 6 6
5 3 3 4 2 3 3 2 4 6 6 4 1 1 1 1 6 6 4 1 6 1 5 6 6 5 6 6 1
2 6 2 1 6 2 5 2 5 3 5 1 6 1 6 5 3 5 6 5 4 5 5 2 2 5 4 5 6
3 2 6 5 5 3 5 1 6 3 3 5 3 1 3 5 5 3 1 1 5 5 5 5 1 6 5 3 6
6 5 5 6 1 5 6 6 5 4 6 6 3 3 4 4 1 2 3 6 3 5 6 6 5 2 6 3 3
4 2 2 5 2 4 3 2 4 6 6 3 1 2 1 6 6 1 4 6 1 2 3 6 2 5 2 6 4
2 2 3 5 5 2 1 6 6 5 5 5 2 5 5 3 6 4 3 6 5 2 1 6 1 5 1 4 3
1 4 3 6 6 6 3 1 4 1 2 6 2 3 3 4 3 5 3 5 5 1 6 1 4 3 4 6 4
5 4 6 6 2 6 2 3 6 2 2 4 5 3 4 3 6 4 1 1 6 4 3 1 3 5 2 6 6
1 3 3 4 3 2 2 5 4 4 1 4 1 1 4 1 5 2 1 2 6 5 2 6 1 5 5 5 3
4 3 6 1 2 1 2 4 4 6 2 5 5 6 1 4 3 3 4 2 1 3 1 4 3 6 2 5 6
6 3 3 2 1 5 3 1 3 1 4 1 6 3 5 4 1 6 3 6 6 5 2 6 6 1 1 5 4
2 6 4 5 3 4 3 4 6 4 6 2 6 6 3 1 3 6 5 6 3 2 5 4 4 4 5 2 6
3 2 5 4 1 3 5 6 2 2 1 1 6 6 4 6 2 3 5 5 2 6 4 6 4 5 3 1 1
1 2 3 2 1 6 3 3 2 1 5 3 1 4 1 5 5 6 5 1 6 3 1 6 5 6 1 5 1
6 6 2 6 2 4 3 6 5 4 4 6 2 3 6 6 6 6 5 1 2

The summary by frequency graph is shown in Figure 2-5.

Summarizing by a frequency chart, gives

Number Showing on Die	1	2	3	4	5	6
Frequency	90	87	98	86	107	132

The summary by frequency graph is shown in Figure 2-6.

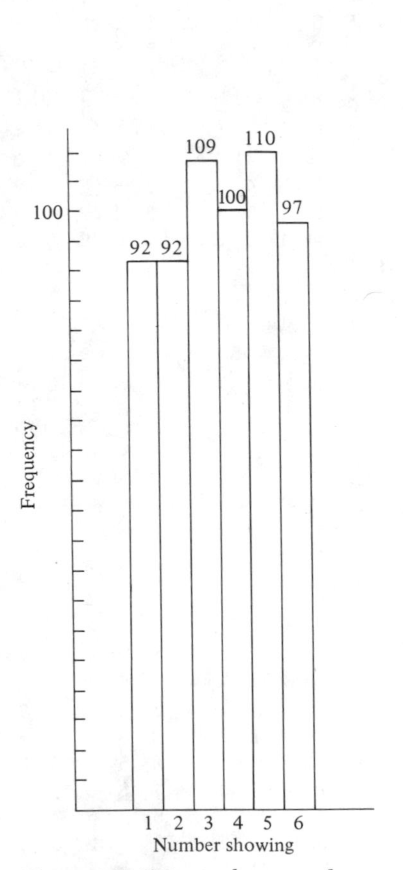

Figure 2-5 600 actual tosses of a fair die.

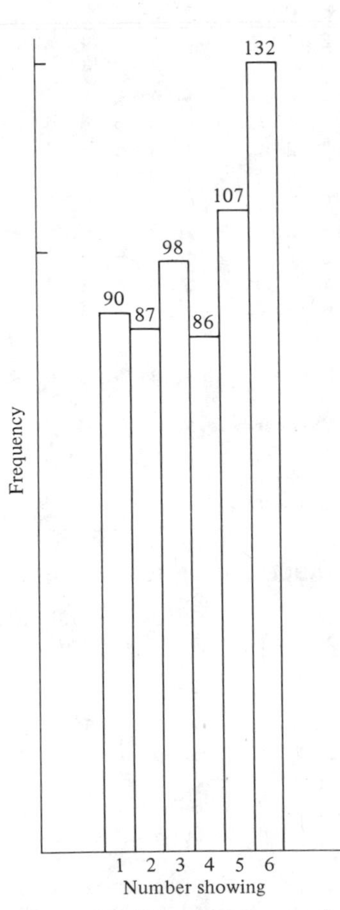

Figure 2-6 600 actual tosses of a biased die.

CLASS EXERCISE

2-18 Can you tell that the biased die *is* biased on the basis of 60 tosses?

____

2-19 Can you tell on the basis of 600 tosses? ________________________

2-20 How would you try to prove that the die was not behaving randomly? ______

It should be clear from this experiment that the smaller your sample is, the easier it is for random fluctuations to obscure any underlying pattern. With a large sample it is usually much easier to spot patterns in the results.

CLASS EXERCISE

In the next three exercises we ask you to answer the questions without pausing to reflect. Put down the very first thing that comes into your head. Are you ready?

2-21 Quickly write down a color. ______

2-22 A number. ______

2-23 A letter of the alphabet. ______

2-24 Analyze the class results. Do you think the selections were random? ______

2-25 Using these results, if you now tried to predict how people will respond to these three questions, do you think that you would be correct more than half of the time? ______

2-26 Do you have ESP? ______

Chapter Summary

We have seen that it is important for samples to be chosen randomly from the correct populations if valid inferences are to be made. We have seen that large samples are less likely to be influenced by random fluctuations than are small samples.

Words You Should Know

Random sample	Frequency graph
Inference	Frequency table

ASSIGNMENT

A sample of ladies by night
Gave Mr. Gallup a fright
When he questioned by day
The answers all lay
Where he'd expected they might.

1 Explain why the commercial quoted at the beginning of this chapter is deceitful advertising.

2 a Suppose you had a random sample of students at this school. If the samples were truly random by age and by sex, then we would be confident that the average age of our sample would be close to the average age of the entire population and that the proportion of males in our sample would be near the proportion of males in the population. Your assignment, if you decide to accept it, is to devise a method of obtaining a sample of 100 students so that the age and sex will be random. Outline this method in a clear, detailed, and specific paragraph. Include the exact questions you will ask. Comment on some of the strengths and some of the weaknesses of your method.

b After your method has been approved by your instructor, gather this data.* Include in your report the data, the average age of your sample, the number of males and the number of females in your sample, and comment on anything that occurred that was not expected. Do you think that the average age of all the students is close to the average you computed for your sample? Do you think that the proportion of males on campus is close to the proportion that you picked?

c Save a copy of the data that you collected for use in later chapters of this book. The oldest person was ________. The youngest was ________. The average age correct to two decimal places: $m =$ ________. The percentage of males ________%. The percentage of females ________%.

List the Ages of the Females	List the Ages of the Males

3 Suggest procedures for getting a random sample to estimate:

a The percentage of students on your campus who own an American-made car.

b The percentage of automobiles in your town which are station wagons.

c The number of days that it will rain in your town during the next calendar year.

d The number of babies that will be born in your town during the next calendar year.

e The number of six-packs of beer that will be sold by a local beer store during the next month.

4 Take five cards numbered from 1 to 5. Shuffle them and arrange them in a row face down. Pick one at random. If it is the 4, you win the game. According to statistical theory, you should win this game about 20 percent of the time. Play

* *Note to the Instructor:* This data should be collected by you for use in Chapter 7.

the game 50 times. Count your wins and calculate your winning percentage by dividing the number of wins by 50. Do you think your percentage is close to the theoretical 20 percent? If not, why not?

5 a If any organization in your town runs a bingo game, visit it one time, and report on the procedure they use to assure randomness in the game. Can the organization possibly lose money? Why?

b A bingo game possibly can end after only four numbers are drawn. It may last much longer, until over 60 numbers are called. Attend a bingo game and record the number of numbers called to win each game. Find the average number of calls. If different members of the class attend different games, are their resulting averages approximately the same?

6 Many times when a speaker wants to illustrate some point, he will give several examples which he claims to have "taken at random" from a large collection of examples. In fact, most of the time, they were *not* chosen at random at all. Just the opposite happened. He probably selected the example very carefully, considering each one and choosing not to use it if he did not like it. Taking examples "at random" should mean that every potential example has an equal chance to be picked. It is not random if certain examples are *sure* to be picked, while others are sure *not* to be picked. Find an example of such a biased sample.

7 Read the introduction and Chapter 17 in M. J. Slonim's little book, "Sampling in a Nutshell," Simon and Schuster, New York, 1966. This is one of the funniest and most delightful books you can find about statistics. Chapter 17 is a collection of case histories ranging from betting at the race track to doing inventory at Sears Roebuck. [*Note:* This book was also issued with the title "Sampling, A Quick Reliable Guide to the Unknown."]

8 Read pages 19–20 in D. Huff, "How to Lie With Statistics," Norton, New York, 1954, which describes how a magazine poll before a U.S. presidential election turned out to be disastrously biased.

Readings from J. Tanur (ed.), "Statistics: A Guide to the Unknown," Holden-Day, San Francisco, 1972.

9 Election Night on Television, Link, p. 137. How did NBC get its sample of returns?

10 Information for the Nation from a Sample Survey, Taeuber, p. 285. How does the government know what the national unemployment rate is?

11 Opinion Polling in a Democracy, Gallup, p. 146. How can Gallup keep track of constantly changing opinion? What if the person has never heard of the topic you want his opinion on?

12 The Plight of the Whales, Chapman, p. 84. How do you get a random sample of blue whales?

Readings from The New York Times

13 Read about using the telephone to sample a population, May 5, 1974, Section 3 p. 15.

THREE

Probability or Chance

A fool must now and then be right by chance.
William Cowper

They are entirely fortuitous you say? Come! Come! Do you really mean that? . . . When the four dice produce the venus throw you may talk of accident: but suppose you made a hundred casts and the venus-throw appeared a hundred times; could you call that accidental?
*Cicero, De Divinatione**
Cicero died in 43 BC

* Quoted from F. N. David, "Games, Gods, and Gambling," Charles Griffin and Company, London, 1971.

INTRODUCTION – TEST YOUR INTUITION

The probability of getting heads on a fair coin is 1/2. Mathematicians measure probability with a number between 0 and 1. In this chapter we will see how to answer questions like these:

What is the probability of tossing tails with a fair coin?
What is the probability of rolling a five on a fair die?
What is the probability of rolling a sum of 5 on 2 fair dice?

To save time, we symbolize these questions as follows:

P(tails on fair coin) = ________
P(five on a fair die) = ________
P(5 on 2 fair dice) = ________

Statisticians use the symbol $P($ $)$ to signify "the probability that." Thus, the probability that a thumbtack will land point up can be written P(a thumbtack lands point up) or more simply P(up).

CLASS EXERCISE

3-1 Guess at the answers to the three questions above. ________, ________, ________.

3-2 *Read* the following six-question test. Do *not* answer the questions yet.

Test

I F. Haxo was (a) a Belgian impressionistic painter, (b) an American Civil War financer, (c) a French military engineer.
II Calluna Vulgari is the Latin name for (a) Lilac, (b) Heather, (c) Rose Fern.
III Esaias Tegner is (a) a Norwegian astronomer, (b) a Swedish poet, (c) a Danish inventor.
IV Guayas is located in (a) Ecuador, (b) Columbia, (c) Cuba.
V Robert Fulton died on February (a) 22d, (b) 23d, (c) 24th.
VI Alofi is in (a) the Wallis Islands, (b) the Futuna Islands, (c) New Caledonia.

Now, *before* you answer the test questions try to answer *this* set of questions. Guess at the answers if you have to.

3-3 Let S stand for the number of questions you will get right. Predict the value of S. ________

3-4 Estimate what fraction of the class will get Question V right. ________

3-5 Estimate P(you get Question V right). ________

3-6 Would you change your answers to Exercises 3-4 and 3-5 if they were about one of the other questions instead of Question V. ________

3-7 Estimate what fraction of the class will get exactly two questions right.

____

3-8 Estimate $P(S = 2)$. _______________

3-9 Estimate $P(S = 0)$. _______________

3-10 Predict the highest score in the class. _______________

3-11 Now answer the six-test questions. Do not leave any blank. Record your answers here: I _______ II _______ III _______ IV _______ V _______ VI _______

3-12 Now correct the tests* and record the number you got right. $S =$ _______

3-13 Let n stand for the number of students who took the test. $n =$ _______

3-14 Compute the fraction of the class that got Question V right. _______

3-15 Compute the fraction of the class that got exactly two questions right.

____

3-16 Record the highest score in the class. _______________

3-17 Why might the actual values in Exercises 3-14, 3-15, and 3-16 be different from what you predicted in Exercises 3-4, 3-7, and 3-10? _______________

We are going to summarize the test results from Exercise 3-2 in Table 3-1.

3-18 Let n equal the number of people who took the test. $n =$ _______

3-19 Multiply n by the first percentage in column A in Table 3-1 (namely .1) and divide by 100. This will give you the number of people who, according to statistical theory, would be expected to get all six test questions right. Record this result at the top of column B.

TABLE 3-1

		Number of People	
S Equals Number Correct	Approximate Percent Expected A	Theoretical B	Actual C
6	.1		
5	1.6		
4	8.2		
3	21.9		
2	32.9		
1	26.3		
0	8.8		

* Correct answers are given on the last page of this chapter.

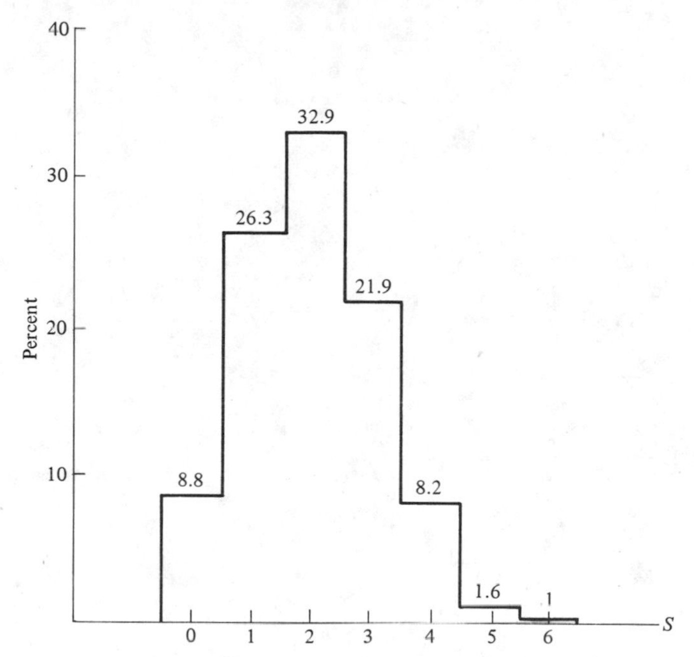

Figure 3-1 Theoretical outcomes for quiz results.

3-20 Record the *actual* number of people who got six right at the top of column *C*. Repeat this whole procedure for the rest of column *B* and column *C*.

3-21 How can you account for any difference between columns *B* and *C* in Table 3-1?

3-22 Graph column *C* on the axes in Figure 3-2.

3-23 Is the shape of your graph close to the theoretical graph in Figure 3-1?

3-24 How can you account for differences that may occur?

Now let us consider the roster for this class.

3-25 What is the probability that the thirteenth person on the alphabetical roster of this class is female? ______________________

3-26 What is the probability that this person is right-handed? ____________

3-27 You all have a nine-digit social security number of the form *aaa-bc-dddd*. If *c* is the middle digit, what is the probability that *c* is odd for this thirteenth person? ______________________

3-28 Guess what percentage of this class will have an odd number for the middle digit:

a Less than 40 percent

b Between 40 and 60 percent

c More than 60 percent

______________________________________.

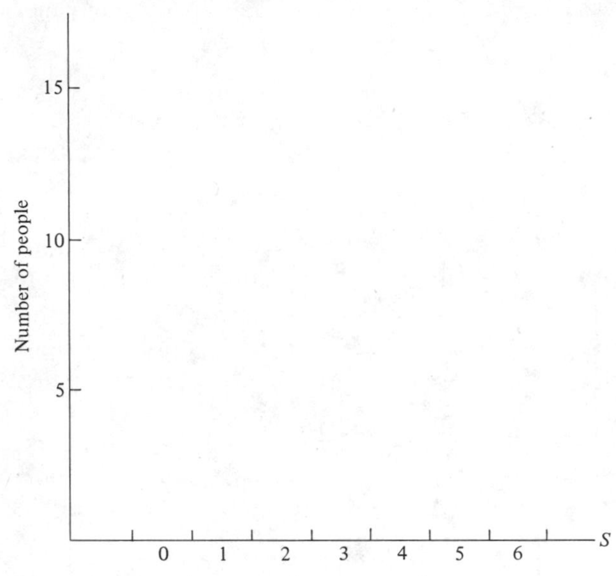

Figure 3-2 Actual outcome for quiz results.

3-29 After you have answered this last question, determine the middle digit of the thirteenth person's number. ____________

3-30 Was it odd or even? ____________

3-31 How many students in this class of the same sex have a *similar* middle digit? ____________

3-32 What conclusions can you draw? ____________

3-33 Count how many persons have odd middle digits in their social security numbers. ____________

3-34 How many are even? ____________

3-35 What percentage have odd numbers in the middle? ____________

3-36 If your answer is not what you predicted in Exercise 3-28 above, can you explain why?

3-37 Guess the probability that two persons in this room have the same birthday, month and day.

- a Less than .4
- b Between .4 and .6
- c Greater than .6

3-38 Would you be willing to wager money that two persons do or do not have the same birthday? ____________

3-39 Have each person write his or her birthday, month and day, on a slip of paper and collect them.

3-40 Are two the same? ______________________________

HOW TO CALCULATE PROBABILITIES

To find the probability of a particular outcome of an experiment, we first consider all the possible outcomes of the experiment. If all of these possible outcomes are *equally likely* to occur, then the probability of the particular outcome of interest—which we shall call the **event**—is equal to the number of favorable outcomes divided by the total number of equally likely outcomes. Recall Exercise 3-27 about the middle digit of social security numbers. Many people mistakenly assume that the probability of an odd middle digit is 1/2 since there are 5 odd digits out of the 10 digits. However, these 10 outcomes are *not* equally likely.

Examples

1 If we roll one die, there are six equally likely outcomes, namely 1, 2, 3, 4, 5, 6. Suppose we let X stand for the outcomes, and then calculate the probability that X is even. Since 3 of the 6 possible outcomes are favorable, that is even, then $P(X$ is even$) = 3/6$ or $1/2$ or $.5$. (*Note:* Any one of these three ways to write the answer is acceptable.) Similarly $P(X$ is greater than $4) = 2/6 = 1/3$ since 2 of the 6 possible outcomes are favorable, that is, greater than 4.

2 If we draw one card from an ordinary deck and we let X stand for the outcome, then $P(X$ is a club$) = 13/52 = 1/4 = .25$. $P(X$ is a jack$) = 4/52 = 1/13 = .08$. $P(X$ is the jack of clubs$) = 1/52 = .02$.

TO CALCULATE THE PROBABILITY OF A PARTICULAR OUTCOME OF AN EXPERIMENT

1 Describe the experiment clearly.

2 State clearly what the particular event is.

3 See that all the possible outcomes of the experiment are *equally likely*. Then count the number of these possible outcomes. Call this n (for number).

4 Count all the outcomes which are favorable to the event. Call this f.

5 Divide f by n. That is,

$$P(\text{event}) = \frac{f}{n}$$

CLASS EXERCISE

3-41 A fair die is tossed. Let X stand for the number that lands on top. Find

a $P(X \text{ is odd})$ ________

b $P(X < 6)$ ________

c $P(X > 0)$ ________

d $P(X > 10)$ ________

3-42 A membership list for a civic club contains 200 names. Of these, 120 are names of females. Let X stand for the sex of each person on the list. If one name is picked at random from the list, find $P(X \text{ is male})$. ________

In some experiments where there are only *two* possible outcomes (such as male-female or odd-even), we arbitrarily call one of the possible outcomes "success" and the other one "failure." We often let lowercase p stand for the probability of "success," and we use q to symbolize the probability of "failure."

Examples

1 If $p = P(X \text{ is female})$, then $q = P(X \text{ is male})$.
2 If $p = 8/9$, then $q = 1/9$.
3 If $p = P(X \text{ has blue eyes})$, then $q = P(X \text{ does not have blue eyes})$.

CLASS EXERCISE

3-43 Suppose $p = P(X \text{ is red}) = .30$. Define $q = P($______$)$

3-44 What is the value of q? ________

3-45 Find the value of $p + q$ ________

3-46 Is this always true? ________

3-47 Express q in terms of p. $q =$ ________

3-48 If $p = P(X \text{ is left handed})$, then $q = P(X$ is ________ $)$

3-49 If $p = .05$, then $q =$ ________

3-50 If $q = .5$, then $p =$ ______________________________

3-51 If $P(X$ wins a chess match$) = .3$ and $P(X$ loses the match$) = .6$, explain why they do not add up to 1. ______________________________

3-52 If $p = P(X$ wins a baseball game), what does q stand for? $q = P($______$)$

3-53 If $p = P(X$ wins a football game), what does q stand for? $q = P($______$)$

3-54 In Exercise 3-53 if $p = .32$, what is the numerical value of q? $q =$ __________

If we toss a penny and a nickel together and we ask how many heads we get, there are three possible outcomes: 2 heads, 1 head, or no heads. But these three outcomes are *not* equally likely, because we are *more* likely to get 1 head than 2 heads. These three outcomes are therefore not helpful to us in calculating probabilities. However, there is a way to look at the experiment so that we can list *four equally likely* outcomes. Each outcome is a *pair*. They are

Outcome	Penny	Nickel
1	Heads	Heads
2	Heads	Tails
3	Tails	Heads
4	Tails	Tails

Therefore, $P(2 \text{ heads}) = 1/4$ since only outcome 1 is favorable, and $P(\text{at least 1 tail}) = 3/4$ since outcomes 2, 3, and 4 are favorable.

CLASS EXERCISE

If a penny, a nickel, and a dime are tossed, there are eight equally likely outcomes. Each outcome consists of three words.

3-55 List them in the chart below:

Outcome	Penny	Nickel	Dime
1	Heads	Heads	Heads
2			
3			
4			
5			
6			
7			
8			

3-56 Let X equal the number of heads when we toss the three coins together. Find:

a $P(X = 2)$ ________ b $P(X \geq 1)$ ________ c $P(X < 2)$ ________

COMPUTER SIMULATION

What does it mean when we say that P(heads) for a fair coin equals 1/2? Roughly speaking, it means that the coin will show heads *about* half the time it is tossed. But you know that it is possible to throw a fair coin, say, 6 times and get almost all heads. In fact, that is not so unusual.

When statisticians say that $P(\text{heads}) = 1/2$, they mean that *in the long run* the proportion of times the coin shows heads will get very close to 1/2. We used the computer to simulate repeated tossing of a fair coin. The computer "tossed the coin" altogether 100,000 times and printed the proportion of heads as it went along. The results are shown in Table 3-2.

You will notice a good bit of fluctuation in the beginning, but as the number of tosses got higher and higher, the percentage of heads got very close to 50 percent and stayed close to 50 percent.

TABLE 3-2 RESULTS OF SIMULATED COIN TOSSING

Number of Tosses	Number of Heads	Percentage of Heads	Number of Tosses	Number of Heads	Percentage of Heads
1	0	0	600	314	52.3333
2	1	50	700	364	52
3	1	33.3333	800	416	52
4	2	50	900	469	52.1111
5	2	40	1,000	514	51.4
6	3	50	2,000	1,005	50.25
7	4	57.1429	3,000	1,498	49.9333
8	4	50	4,000	1,999	49.975
9	4	44.4444	5,000	2,498	49.96
10	4	40	6,000	2,991	49.85
20	14	70	7,000	3,453	49.3286
30	20	66.6667	8,000	3,941	49.2625
40	26	65	9,000	4,459	49.5444
50	28	56	10,000	4,944	49.44
60	31	51.6667	20,000	9,994	49.97
70	34	48.5714	30,000	15,036	50.12
80	40	50	40,000	20,092	50.23
90	45	50	50,000	25,058	50.116
100	52	52	60,000	30,034	50.0567
200	112	56	70,000	35,016	50.0229
300	164	54.6667	80,000	39,978	49.9725
400	218	54.5	90,000	44,992	49.9911
500	265	53	100,000	49,909	49.909

CLASS EXERCISE

Suppose each person in this class does the following: Toss a coin in the air. If heads you win, write +1. If tails, you lose, write −1. Now toss the coin again. If heads, add 1 to your score. If tails, subtract 1 from your score. Repeat this procedure 10 times. For instance, if Al, Bob, and Carmine each tossed a coin 10 times, they might have tossed as follows:

Toss	*A*	*B*	*C*
1	H	T	H
2	H	T	T
3	T	T	H
4	H	H	T
5	T	T	H
6	H	H	H
7	T	T	H
8	H	T	H
9	H	T	H
10	H	T	H

3-57 Find the score for Al. ____________________

3-58 Find the score for Bob. ____________________

3-59 Find the score for Carmine. ____________________

You should have gotten 4, −6, +6 as indicated below:

Toss	*A*	*B*	*C*
1	1	−1	1
2	2	−2	0
3	1	−3	1
4	2	2	0
5	1	−3	1
6	2	−2	2
7	1	−3	3
8	2	−4	4
9	3	5	5
10	4	−6	6

3-60 Take a guess and try to estimate what your score would be if you tossed a coin 10 times. ____________________

3-61 What is the highest *possible* score anyone can get? ____________________

3-62 The lowest possible score? ____________________

3-63 If the entire class did this, what do you think the highest score would actually be? __________ (Would you bet money on it? __________)

3-64 What will the lowest actual score be? ____________________

3-65 What will the class average be if everyone in this class tosses a coin 10 times? __________ Which score do you think will be the most common? __________ (This would be called the **mode** of all the scores.)

3-66 What percentage of the class will probably end up with a positive score? __________ Why is this a tricky question? ______________________________

3-67 Now (or tonight at home) have each person toss a coin ten times. Record your answers below.

Toss	H or T	Score
1		
2		
3		
4		
5		
6		
7		
8		
9		
10		

3-68 What was your score? ______________________________

3-69 Record the scores of each person in the class in the chart below.

Score	Frequency
10	
8	
6	
4	
2	
0	
−2	
−4	
−6	
−8	
−10	
Total	$n =$

In theory we expect:

Score	10	8	6	4	2	0	−2	−4	−6	−8	−10
Percent	0.1	1.0	4.4	11.7	20.5	24.6	20.5	11.7	4.4	1.0	0.1

The theoretical mean is 0. Therefore, 0 is the best estimate of an individual score. We expect about 98 percent of the class to score between −6 and +6 inclusive. About 89 percent of the class should score between −4 and +4 inclusive.

ESTIMATING POPULATION PERCENTAGES

Suppose we wish to estimate the percentage of pay telephones in New Jersey gas stations, which are out of order. There are three ways to look at this one idea. We can say that we are estimating:

1 The percentage of pay phones out of order.
2 The proportion of pay phones out of order.
3 The probability that a pay phone picked at random is out of order.

We do not know the true value of this probability, and if we use data from a random sample, we will have only an *estimate* of the true value. Statisticians frequently use English letters for such estimates and Greek letters for the true value. Hence, we could use the English p for our estimate and the corresponding Greek π (pi) for the true value. Note that this use of the symbol π does not stand for 3.14 · · · , but rather for the true value of a probability in some population.

CLASS EXERCISE

A random sample of 400 pay phones showed that 120 were out of order.

3-70 Define π. ______

3-71 Define p. ______

3-72 Define q. ______

3-73 What is the numerical value of p? ______

3-74 What is the numerical value of π? ______

3-75 What is the numerical value of q? ______

Chapter Summary

1 We have seen that in a random experiment with equally likely outcomes, we can calculate the probability of any particular outcome or set of outcomes by using the formula

$$P(\text{event}) = \frac{f}{n}$$

2 We have seen that probability of an event is the same number as the proportion of times that event occurs out of all possible outcomes when an experiment is repeated many times.

Words and Symbols You Should Know

Equally likely outcomes	Success and failure	P(event)
Event	f	p
Probability	n	q

ASSIGNMENT

An old statistician from Trent
Whose brains were irreparably bent
Counted the males
Amongst all the whales
And got more than a hundred percent

1 The first game of a Dodger-Yankee World Series has two possible outcomes. Hence, P(Dodgers win) = 1/2. Comment.

2 a Suppose 20 cards are numbered 1, 2, . . . , 20 and are then mixed. If one is drawn at random, what is the probability that you could guess which one? Would you care to wager some money that you could guess correctly? Would you take the bet if somebody else said that they could do it?
 b Same circumstances as part a but with 100 cards.
 c Is it *possible* to shuffle 20 cards and correctly *guess* the top card?
 d If someone did this, would you attribute it to good luck?
 e If you had wagered $5 that he would fail to name the card and you lost, would you attribute it to good luck?
 f If someone correctly named the card and you stated that therefore the game was not random, but rather that some trick was involved, what is the probability that you are wrong and that it really was just a lucky guess?

3 The long sides of two triangular rods are numbered 1, 2, and 3. They are tossed out at random and the outcome is the side facing down. There are nine possible outcomes.

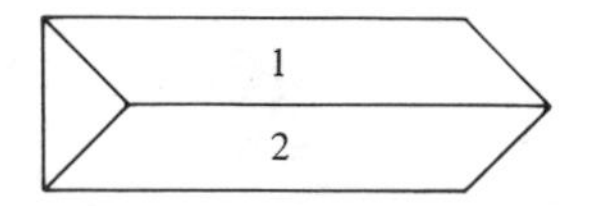

 a List them below.

Outcome	First Rod	Second Rod
1		
2		
3		
4		
5		
6		
7		
8		
9		

b If the outcomes on the two rods are added and their sum is denoted by X, then X can be 2, 3, 4, 5, or 6. Find $P(X = 4)$.

c Find $P(X \text{ is greater than } 4)$.

4 At a neighborhood dance given by the Society for the Abolition of Speling Tests, 365 raffle tickets were sold. The tickets were numbered from 1 to 365.

a Sean bought ticket number 63. What is the probability that he will win?

b Ivan bought a block of tickets from 12 to 22. What is the probability that he will win?

c Jean has ticket 17. What is the probability that he will lose?

d What is the probability that an odd-numbered ticket will be the winning one?

e John has a ticket numbered 400. What is the probability that John will win?

f What is the probability that a ticket with a number that is evenly divisible by 6 will win?

5 In a random sample of 60 biology majors at Sexist University, 40 were female. Consider the question: What percentage of all biology majors are female?

a Define π.

b Define p.

c What is the value of π?

d What is the value of p?

e What is the value of q?

6 a Get 10 identical thumbtacks. Toss them in the air. Record how many land point up. Toss them again and again. Each time record the number of tacks which land point up. Record the cummulative percentages of thumbtacks up. Estimate the probability that one such tack will land up.

The results of one actual experiment were as follows:

Toss Number	Number Landing Up	Cumulative Number Landing Up	Cumulative Percentage Up
1	7	7	70
2	5	12	60
3	4	16	53
4	5	21	50
5	5	26	52
6	4	30	50
7	7	37	53
8	4	41	51
9	6	47	52
10	5	52	52
11	7	59	54
12	5	64	53
13	6	70	54
14	8	78	56
15	2	80	53

b If $\pi = P(\text{a thumbtack lands up})$, then $p = .53$. Estimate a range of values in which you have confidence that the true value of π lies. (*Note to the Instructor: Students can either gather their own data or use our figures. Data should be saved for Assignment 29 in Chapter 9.*)

7 If you guess randomly at four true-false questions, then there are 16 equally likely possible outcomes.

a Using *g* for good guess and *b* for *b*ad guess list the 16 possible outcomes.

b If X equals the number of good guesses that you make when you guess at the four questions, find $P(X = 4)$.

c Suppose a classmate claims to have guessed all four answers correctly. Would you believe her claim? (Reread the quote that opened this chapter.)

d If you do not believe her, what is the probability that you are wrong in rejecting her claim?

8 Why would you not use the answer to Question 7a if the four questions had been multiple-choice questions with three choices a, b, or c for each answer instead of four true-false questions?

9 Explain the meaning of each of these statements. They were all taken from published reports. How do you think each statement was arrived at?

a The probability that a black American carries sickle cell anemia is 6.7%.

b The probability that a Jewish American couple have the genetic background to produce a child with Tay-Sachs disease is .001.

c The probability that you will win money by playing Craps is .49.

d 1/10 Americans suffer from mental illness.

e The probability that a marriage between two Americans will end in divorce is .30.

f The probability that an I.U.D. birth control device will prevent conception is .98.

g The probability of rain tomorrow is .20.

h The probability that an entering college freshman will drop out is .47.

10 Explain each of these:

a How can a television network announce the winner of an election when only a small percentage of the votes have been counted?

b How can the Traffic Safety Council predict 2 weeks in advance the number of automobile accidents that will occur on a holiday?

c How can the Admissions Office of your college predict the number of accepted students who will actually decide to attend the college?

d For parts a, b, and c, describe conditions which might occur which would throw these predictions off.

11 Probabilities are numbers.

a What is the biggest number a probability can be?

b What is the smallest number a probability can be?

12 Make up your own question and use the data in Appendix A to answer it. Use such questions as: What is the probability that a female at Nassau Community College is Catholic? What is the probability that a Liberal has a cumulative average 3.0 or higher? Etc.

13 Read Inflexible Logic by Russell Maloney, which appears in Clifton Fadiman (ed.), "Fantasia Mathematica," Simon and Schuster, New York, 1953. This is a funny short story about how one rich man reacted when he heard that six

monkeys picking randomly at six typewriters would eventually type out exact copies of all the books in the British Museum.

14 Read Chapter 1 in F. N. David, "Games, Gods and Gambling," Charles Griffin and Company, London, 1971. The chapter deals with ancient gambling devices and games and includes photographs of ancient game boards and "dice."

15 Read Darrel Huff's short, light book on probability, "How to Take a Chance," Norton, New York, 1959.

Readings from J. Tanur (ed.), "Statistics: A Guide to the Unknown," Holden-Day, San Francisco, 1972.

16 The Probability of Rain, Miller, p. 372. What does "30% chance of rain" mean?

17 Parking Tickets and Missing Women, Zeisel and Kalven, p. 102. How to beat a parking ticket. The case of Dr. Spock.

18 Statistics, Sports and Some Other Things, Hooke, p. 244. Should you bunt with no out and a man on first?

19 Statistics and Probability Applied to Antiaircraft Fire in World War II, Pearson, p. 407. What does probability have to do with weapon design? How does the statistician function as part of a research team?

20 The Use of Subjective Probability Methods in Estimating Demand, Schwartz, p. 212. What is "subjective probability"?

Answers to test in the beginning of the chapter:

I. c   II. b   III. b   IV. a   V. c   VI. b

FOUR

Did the Expected Thing Happen? Chi Square

It is truth very certain that, when it is not in our power to determine what is true, we ought to follow what is most probable.

René Descartes

Margie May Kabuck went to celebrate the Festival of St. Anthony in Manhattan. After a while she stopped at a table where a man was rolling a giant-sized die. Being a serious gambler, she carefully observed the game for 300 rolls of the die. The results are given in Table 4-1.

TABLE 4-1 TABLE OF OBSERVED VALUES

Number on Die	1	2	3	4	5	6	
Number of times it was observed	60	100	40	47	10	43	Total = 300

She decided that the giant die did not behave like a fair die. Why? Because she knew what kind of results to expect from a fair die and the results she actually observed seemed quite different to her. She *expected* the results to be very close to the theoretically perfect ones given in Table 4-2.

TABLE 4-2 TABLE OF EXPECTED VALUES

Number on Die	1	2	3	4	5	6	
Number of times it was expected	50	50	50	50	50	50	Total = 300

Margie felt intuitively that the carnival die was different from a fair die. In many experiments statisticians must decide whether the observed results are significantly different from the results expected according to some theory. (In this case the theory is that the die is fair.) If the observed results *are different enough*, the statistician will decide that this theory is *not* the correct explanation for the observed results and will search for some other explanation.

In this chapter we are going to illustrate one popular method statisticians use to compare actual *observed* results of an experiment with the results *expected* according to some theory. The general procedure is this:

1 State the theory clearly. What is the hypothesis that you are about to test?
2 Record the observed values.
3 Compute the expected values.
4 Use these two sets of results to calculate a new number called a **test statistic** (denoted by X^2).
5 This statistic measures the difference between the observed results and the theoretical results. If this difference is small, the value of X^2 will be small and the theory might well be cor-

rect. However if the value of X^2 is large, the theory is probably *not* correct.

There are two technical ideas involved when we compute X^2. The first is how to compute it, and the second is how to decide if it is "large."

How to Compute X^2

1 Record the *observed* values (O)

1	2	3	4	5	6	
60	100	40	47	10	43	Total: $n = 300$

2 Record the *expected* values (E)

1	2	3	4	5	6	
50	50	50	50	50	50	Total: $n = 300$

3 Square each value of O and divide by the corresponding E. Then add all these together:

$$\frac{60^2}{50} + \frac{100^2}{50} + \frac{40^2}{50} + \frac{47^2}{50} + \frac{10^2}{50} + \frac{43^2}{50}$$

$$= \frac{3{,}600}{50} + \frac{10{,}000}{50} + \frac{1{,}600}{50} + \frac{2{,}209}{50} + \frac{100}{50} + \frac{1{,}849}{50}$$

$$= 72 + 200 + 32 + 44.18 + 2 + 36.98$$

$$= 387.16$$

4 Subtract the *total number* of outcomes (n)

$$X^2 = 387.16 - 300 = \underline{87.16}$$

Notice that 387.16 is greater than 300 and so X^2 is positive. X^2 is always positive or zero. If your calculations produce a negative value for X^2, you have made a mistake.

Formula for X^2

We use the Greek letter sigma, Σ, to mean "find the sum of a set of numbers." In finding X^2, we summed the values of O^2/E and then subtracted n. We can summarize this by the formula

$$X^2 = \sum \frac{O^2}{E} - n$$

CLASS EXERCISE

A die was rolled 120 times, here are the results:

Number on die	1	2	3	4	5	6
Number of times it was observed	18	16	20	30	17	19

4-1 What is the value of n? ______

4-2 What are the expected values for 120 tosses of a fair die? ______, ______, ______, ______, ______, ______

4-3 Guess whether the value of X^2 in this problem is going to be smaller or larger than the 87.16 in the illustration above. ______

4-4 Compute X^2. ______

We said above that there were two technical ideas in computing X^2. The first was the formula itself. The second is how to decide if the computed value of X^2 is large enough for you to reject the original theory (from which you got your expected values). The value of X^2 should cause you to reject the theory if it is a number higher than a value listed in what statisticians call a **chi-square table.** Chi is the Greek letter equivalent to the English letter X.

To read the chi-square table you must know a number called the **degrees of freedom** (df) for your experiment. We will explain this shortly. For now we will simply tell you that in order to find the degrees of freedom in this dice experiment you subtract one from the number of columns in the table of results. In this case the degrees of freedom equal $6 - 1 = 5$.

A partial chi-square table is printed in Table 4-3. The two col-

TABLE 4-3 CHI-SQUARE TABLE

df	X_c^2
1	6.635
2	9.21
3	11.34
4	13.28
5	15.09
6	16.81

umns are labeled df for "degrees of freedom" and X_c^2 for "critical value of X^2." Since there are 5 degrees of freedom in this problem, $X_c^2 = 15.09$.

What does this 15.09 mean? If we rolled a fair die 300 times and got exactly 50 ones, 50 twos, 50 threes, 50 fours, 50 fives, and 50 sixes, then the value of X^2 would be exactly 0. We are not surprised at slight variations in the outcomes; for example, 52 ones, 49 twos, etc., would not be unusual. In this case the value of X^2 would be slightly more than 0. Statisticians have figured out that the value of X^2 for a dice experiment when the die is fair will almost always be smaller than 15.09. This means that if many persons rolled a *fair* die a large number of times and each person computed a value of X^2, then almost all of them would have values of X^2 less than 15.09. If you roll a die that may or may not be fair and you compute X^2 and get a number larger than 15.09, you can be almost positive that this is due to the fact that the die is not fair. In the particular experiment above, we computed X^2 to be 87.16. Hence we conclude the die is probably not fair.

CLASS EXERCISE

In Exercise 4-4 you computed a value of X^2. The correct value is 6.5.

4-5 How many degrees of freedom were there in this experiment? 5

4-6 What is the appropriate value of X_c^2 in the table? ______

4-7 Do you conclude that this die is probably not fair? ______

There are two questions that you will probably have at this point. First, how did the statisticians decide what formula to use for this type of experiment? Second, how did statisticians arrive at the numbers in the table? To answer the first question: this formula was first suggested about 70 years ago by Karl Pearson, a very inventive statistician from England, who showed by mathematical discussions too advanced for this book that the formula was reasonable. If you take a more advanced statistics course you will probably study the development of this formula. We will only say here that it essentially compares the two sets of results, the observed and expected, entry by entry and computes the overall effect of the individual differences. You will check for yourself in Assignment 2 that if the two sets of results, the observed and the expected, are exactly the same, then your calculated value of X^2 will be 0. As the two sets of results get more and more different, the value of X^2 gets larger and larger.

COMPUTER SIMULATION

For the second question about the entries in the table, we will turn to a computer simulation. Set the computer to simulate a fair die. (Therefore, we know the expected results.) Now roll this computer "die" 300 times and observe the results. (Just as Margie did.) Compute your value of X^2 and record it. Now repeat this whole experiment many times. This will give us a good idea of what values of X^2 we *usually* get with a fair die. Then we will be able to recognize a value of X^2 which is exceptionally high. If we get such a high value in a real experiment with a real die, we will conclude that *probably* the die is not fair.

Printed in Table 4-4 are the results of the first five computer simulations of a fair die.

TABLE 4-4 SIMULATED FAIR DIE TOSS

	Number Showing on Die						
Experiment	**1**	**2**	**3**	**4**	**5**	**6**	X^2
1	48	52	40	54	52	54	2.88
2	48	49	55	62	51	35	8.00
3	62	41	49	57	42	49	6.80
4	56	47	53	46	45	43	3.68
5	40	59	50	57	51	43	5.60

We ran the program 95 more times to get a list of 100 values of X^2. They are printed in Table 4-5, in the order in which they occurred, and then in Table 4-6 they are rearranged and renumbered in numerical order for easier reading. You will notice that all of our experimental results are less than 10. According to mathematical theory if we repeated the experiments many more times, then 95 percent of the time the experiment should produce a value of X^2 less than 11.07, and 99 percent of the time a value less than 15.09. (You can see for yourself that a value of X^2 more than 10 would be unusual.) This means that about 99 percent of the time a fair die gives X^2 less than 15.09. Or conversely, only about 1 percent of the time will X^2 be bigger than 15.09. That in turn means *in terms of probability* that if *you* perform the dice experiment and compute X^2, then if your X^2 is bigger than 15.09, there is at most a 1 percent probability that you have a fair die. In other words, if your X^2 is larger than 15.09, you probably do *not* have a fair die. Therefore, the number 15.09 is entered in the chi-square table as the *critical value* of X^2. We write $X_c^2 = 15.09$.

We have drawn a graph to represent the 100 values of X^2 from the computer experiment. Statisticians often draw graphs when they are trying to represent a lot of numbers. It is usually easier to see if there is any pattern to the numbers by looking at a graph.

Here is how we made the graph. We took the values of X^2 and grouped together those that were close to each other. Then we made a frequency table for these grouped values (see Table 4-7).

Figure 4-1 is a graph of these values. We have drawn a smooth curve through the tops of the bars to get a rough picture of the "shape" or pattern of the distribution of X^2 values. You might guess that if we did another series of 100 simulations and made another graph, then this

TABLE 4-5 RESULTS OF THE COMPUTER SIMULATION

Experiment	X^2	Experiment	X^2	Experiment	X^2
1	2.87	35	7.51	69	4.79
2	7.99	36	5.99	70	3.19
3	6.79	37	4.95	71	4.55
4	3.67	38	3.99	72	9.15
5	5.59	39	3.99	73	3.59
6	5.39	40	4.15	74	5.91
7	1.15	41	.47	75	2.47
8	7.15	42	3.31	76	4.39
9	3.79	43	6.91	77	5.99
10	6.07	44	3.11	78	4.99
11	4.39	45	2.79	79	6.03
12	3.03	46	2.99	80	6.59
13	4.51	47	2.55	81	3.83
14	.59	48	2.15	82	1.51
15	4.03	49	6.91	83	2.91
16	4.43	50	4.23	84	6.91
17	2.51	51	3.23	85	2.91
18	8.43	52	3.03	86	2.59
19	6.39	53	6.23	87	6.19
20	5.87	54	5.35	88	3.99
21	3.83	55	1.59	89	2.43
22	7.55	56	4.71	90	7.35
23	8.35	57	5.87	91	1.59
24	3.35	58	1.39	92	1.47
25	5.07	59	7.23	93	4.59
26	6.07	60	7.31	94	5.11
27	3.71	61	4.19	95	4.47
28	4.15	62	2.03	96	1.95
29	3.55	63	8.43	97	6.63
30	1.31	64	8.79	98	5.19
31	7.47	65	2.47	99	2.43
32	4.83	66	4.55	100	7.83
33	9.03	67	3.63		
34	3.91	68	3.35		

new graph would have just about the same shape as this one. There is a basic fundamental shape that all these experimental graphs would have.

Mathematicians have studied this problem and have established what the "ideal" or "theoretically perfect" shape is for distributions of X^2 values. This theoretical graph is known as the graph of the chi-square distribution. It is from these theoretical graphs that the values in the chi-square table are taken. Figure 4-2 shows a theoretical graph with 5 degrees of freedom.

TABLE 4-6 RESULTS OF THE COMPUTER SIMULATION REARRANGED IN NUMERICAL ORDER

Renumbered Result	X^2	Renumbered Result	X^2	Renumbered Result	X^2
1	.47	35	3.63	69	5.87
2	.59	36	3.67	70	5.87
3	1.15	37	3.71	71	5.91
4	1.31	38	3.79	72	5.99
5	1.39	39	3.83	73	5.99
6	1.47	40	3.83	74	6.03
7	1.51	41	3.91	75	6.07
8	1.59	42	3.99	76	6.07
9	1.59	43	3.99	77	6.19
10	1.95	44	3.99	78	6.23
11	2.03	45	4.03	79	6.39
12	2.15	46	4.15	80	6.59
13	2.43	47	4.15	81	6.63
14	2.43	48	4.19	82	6.79
15	2.47	49	4.23	83	6.91
16	2.47	50	4.29	84	6.91
17	2.51	51	4.39	85	6.91
18	2.55	52	4.43	86	7.15
19	2.59	53	4.47	87	7.23
20	2.79	54	4.51	88	7.31
21	2.87	55	4.55	89	7.35
22	2.91	56	4.55	90	7.47
23	2.91	57	4.59	91	7.51
24	2.99	58	4.71	92	7.55
25	3.03	59	4.79	93	7.83
26	3.03	60	4.83	94	7.99
27	3.11	61	4.95	95	8.35
28	3.19	62	4.99	96	8.43
29	3.23	63	5.07	97	8.43
30	3.31	64	5.11	98	8.79
31	3.35	65	5.19	99	9.03
32	3.35	66	5.35	100	9.15
33	3.55	67	5.39		
34	3.59	68	5.59		

TABLE 4-7 FREQUENCY TABLE

Value of X^2	0–1	1–2	2–3	3–4	4–5	5–6	6–7	7–8	8–9	9–10
Frequency (How Many Were in Each Group?)	2	8	14	20	18	11	12	9	4	2

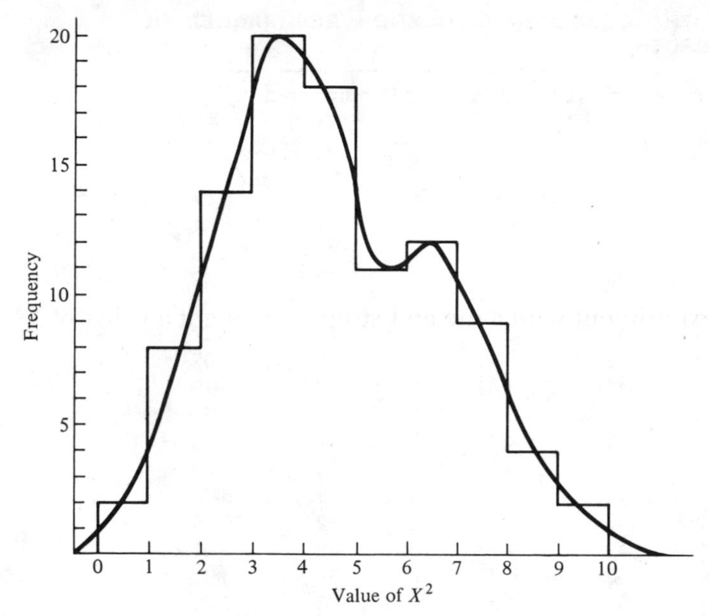

Figure 4-1 X^2 values from a fair die experiment.

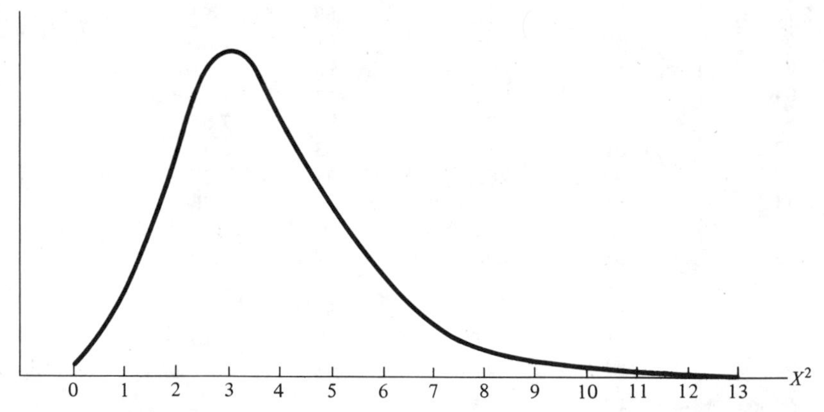

Figure 4-2 Theoretical graph for chi square with 5 degrees of freedom.

Critical Values and Significance Levels

Suppose the die of our experiment is really fair. Statisticians call this assumption of fairness a **null hypothesis.** That means if we *kept on* rolling the die (forever), the results would get closer and closer to perfect (1/6 ones, 1/6 twos, 1/6 threes, 1/6 fours, 1/6 fives, and 1/6 sixes). However, we are only rolling it 300 times and therefore, *even if the die is fair,* we do not expect perfect results. We expect to be off by a little. Therefore, we will not get X^2 exactly equal to zero even for an experiment with a fair die, but we will probably get a small value of X^2 with a fair die.

The graph in Figure 4-2 shows which values of X^2 occur frequently and which do not in experiments with fair dice. Where the graph is high (say around 3 or 4), these are values of X^2 that are quite common. Where the graph is low (as you get past about 10), these are values of X^2 that hardly ever occur with a fair die.

On the graph in Figure 4-2 you can check that 95 percent of these values of X^2 are less than 11.07 and 99 percent are less than 15.09. Consider what this means in terms of probability. Suppose you do an experiment with a die and suppose you get a value of X^2 bigger than 15.09. This means there is only a 1 percent probability that the die is fair. In short, the die is probably not fair.

This means that you will state on the basis of your experiment that you have proved the die is *not* fair. However, you must realize that there is a .01 probability that you are making a mistake. There is a .01 probability that the die actually is fair and that your experiment just by chance turned up a very untypical set of random outcomes. This probability of error is called the **significance level** of your experiment and is denoted by the Greek letter alpha, α. Whenever you reach a conclusion about a statistical hypothesis you should also state the significance level, so that other people will know the probability that your experiment might have led you to a wrong conclusion. It is quite standard for statistical experiments to be carried out at the .01 significance level or the .05 significance level, but other values are occasionally used. The experimenter is free to choose any significance level he wants. However, if he intends to offer his experiment as proof of some idea, then the significance level chosen must be acceptable to other workers in that same field. If α is too large, there is a good chance that the results are not conclusive but possibly just due to chance. That is, if the experiment were repeated a second time, the results might be quite different.

Look at the chi-square table in Appendix B. Next to each value of degrees of freedom we have printed two values of X^2. Under .01 you will find the value of X^2 that splits off the rightmost 1 percent of the area; this is the value you would use for testing at the .01 significance level. Under .05 you will find the value that splits off the rightmost 5 percent of the area; this is for testing at the .05 significance level.

STATISTICAL HYPOTHESES

In Chapter 1 we said that a statistical hypothesis is a statement made about an *entire* population. In a more formal sense now we will say that a statistical hypothesis is a statement about an entire population which we would like to confirm or deny on the basis of information gathered from a random sample of that population. Numerical information based on only a *sample* from a population is called a **statistic.** Therefore, a statistical hypothesis is a statement about a population which we will attempt to confirm or deny by looking at sample statistics.

In the dice problem there are two opposite hypotheses about the population: either the die is fair or the die is biased. If we assume that the die is fair, then we can know what to expect. You recall that Margie May Kabuck expected about 50 ones, 50 twos, etc., for a fair die. If, however, we were to assume that the die is biased, we would not necessarily know what outcomes to expect. Statisticians call the assumption which leads to knowledge of the expected values the **null hypothesis.** This hypothesis is technically the one that is tested. It is denoted by H_0 ("H sub 0"). The other hypothesis, that the die is biased, is called the **alternative hypothesis** and is denoted by H_a ("H sub a").

FORMAL CHI-SQUARE TEST PROCEDURE

Let us go back and examine the chi-square testing procedure more formally now. Once again we use the dice experiment as an illustration. The population is all possible rolls of the dice. The sample is the 300 rolls actually done in the experiment.

1 State the hypothesis

H_0 : The die is fair.
H_a : The die is not fair.

2 Record observed values

Number Showing	1	2	3	4	5	6
Frequency	60	100	40	47	10	43

3 Compute expected values

Number Showing	1	2	3	4	5	6
Frequency	50	50	50	50	50	50

4 Compute X^2:

$X^2 = 87.16$ (by our previous computation)

5 Find tabled value of X_c^2

1 Select significance level. We choose $\alpha = .01$.
2 Compute degrees of freedom. $df = 6 - 1 = 5$
3 Record tabled value. $X_c^2 = 15.09$

6 Formulate a decision rule

We will reject the null hypothesis if our outcome is greater than 15.09. If X^2 is greater than 15.09 in our sample, this means that it is so large that the die is almost certainly not fair.

7 Conclusion

Since $87.16 > 15.09$, we reject the null hypothesis. Therefore, H_0 is probably false and the die is probably not fair.

Important: In any statistical experiment, the size of the sample must be large enough if we wish to make useful inferences about the population. In a chi-square experiment the sample is large enough when the expected values are each greater than 5. (More advanced text books state exceptions where some expected values can be as low as 1.)

If n is too small, there is too much possibility for a few variations to be misleading. See Figures 2-4 and 2-6: the 60 rolls of biased die as compared to the 300 rolls. See also Assignment 26 and 27 at the end of this chapter.

CONCLUSIONS TO A HYPOTHESIS TEST: TYPE I AND TYPE II ERRORS

When you finish a hypothesis test, you will state one of two conclusions: "I reject the null hypothesis," or "I do not reject the null hypothesis." You realize from the discussion earlier in this chapter that even if you have done your experiment carefully and your computations perfectly, there is still some probability that you have come to the wrong conclusion. That is because you are only looking at a sample from the population, and just by luck you may have gotten a very untypical sample. We point out again that these errors are part of the nature of random sampling and that you have not done something wrong.

Therefore at the end of your experiment, after you have announced your conclusion, there are four possible situations which may have resulted.

1 You said, "Reject null hypothesis," and you were right.
2 You said, "Reject null hypothesis," and you were wrong.
3 You said, "Do not reject null hypothesis," and you were right.
4 You said, "Do not reject null hypothesis," and you were wrong.

The statistician is always expected to know and to state the probability that his announced conclusion is wrong. This allows others who read the results of his experiment to know how much "faith" to have in the results.

Let us look more carefully at the two situations which led to wrong conclusions. Those were situations 2 and 4.

In situation 2 you reject the null hypothesis even though it happens to be true. This is called a **Type I error.** For example, you are testing a coin for fairness, and you get 48 out of 50 heads. You decide that the coin is biased. H_0 was "the coin is fair." Your sample leads you to reject H_0 since you now think that the coin gets too many heads. However, it is still possible that the coin *is* fair and that if you had *kept on* tossing it, the percent of heads and tails would have equalized. If that is the case, then your decision was wrong. You committed a Type I error.

When you pick the significance level α for your experiment, you are fixing the probability of making a Type I error. The probability of a Type I error cannot be more than α. In other words, you can control the probability of a Type I error by picking the appropriate significance level.

In situation 4 you said, "do not reject the null hypothesis," even though it happens to be false. In other words the evidence from your sample was what you would expect if the null hypothesis were true. This is called a **Type II error.**

It is not so easy to know what the probability of a Type II error is. We see that it occurs when the null hypothesis is false, but to measure it we need to know precisely which alternative hypothesis is true. For example, we test a biased coin and our sample leads us to decide it is fair. But to compute the probability of such an error, we would need to know just *how biased* the coin is. Often the statistician can only approximate the probability of a Type II error. We will not deal with this topic in the main text of the book, but if you are interested in pursuing it, see Appendix E.

Illustration of the Two Types of Errors

Suppose we have four coins: a penny, a nickel, a dime, and a quarter. For the sake of discussion assume that the first two are biased, and the last two are fair. We test each one by tossing it 100 times. We set the significance level α at .05. Our null hypothesis is "the coin is fair." Here are some possible results:

Coin	Actual Case	Sample Evidence	Conclusion Based on Evidence	Error
Penny	Biased	79 heads	Biased – reject H_0	None
Nickel	Biased	45 heads	Fair – do not reject H_0	Type II
Dime	Fair	70 heads	Biased – reject H_0	Type I
Quarter	Fair	56 heads	Fair – do not reject H_0	None

Setting $\alpha = .05$ means that *if the H_0 is actually true,* we run a 5 percent chance of calling it false and a 95 percent chance of not calling it false. It tells us nothing if the H_0 is actually false.

Our conclusion is based on our sample evidence. Whether or not the sample is *random* depends on the *procedure* that was used to select it. If a fair procedure was used so that each member of the population had the same *chance* to be chosen, then it is random. A random sample *may turn out to be* very unrepresentative of the population. That in itself does not mean there is anything wrong with the procedure used to get the sample. The larger a random sample is, the more likely that it will be representative.

In hypothesis testing the assumption is that the sample *is* a random sample. Then we know that setting the significance level is the way that we make allowances for untypical samples. Setting the significance level does not guard against biased *procedures* for getting the samples. If the experimenter has set up the experiment in such a way that he does not get random samples, then further analysis may be worthless.

Another Illustration of a Chi-square Test

A biology student, Gene Poole, is studying genetics. He is told that for a certain type of insect, one-fourth of the males have red eyes, one-fourth have black eyes, and one-half have gray eyes. Gene goes to much trouble to breed a lot of these insects. He carefully examines them to find the males. He puts all the males in one jar and then he carefully looks at their eyes. After 16 hours of sorting and counting he arrives at these results:

Eye Color	Red	Black	Grey	
Frequency	133	155	312	Total = 600

Are these different enough from the expected results to suspect that the genetic theory does not apply in his population?

SOLUTION

1 State hypothesis

Population: All male insects that would be bred if he continued the experiment.
Sample: The 600 male insects actually bred.
Null hypothesis: H_0 : Genetic theory ($\frac{1}{4}$ red, $\frac{1}{4}$ black, $\frac{1}{2}$ grey) is correct for this population. This is the *null* hypothesis because it leads to the expected values.
Alternative: H_a : Genetic theory does not hold.

2 Observed values

Eye Color	Red	Black	Grey	
Frequency	133	155	312	Total = 600

3 Expected values

Assuming the null hypothesis is true, we expect $\frac{1}{4}$ of the 600 insects to have red eyes, $\frac{1}{4}$ to have black eyes, and $\frac{1}{2}$ to have grey eyes.

$\frac{1}{4}(600) = 150$
$\frac{1}{2}(600) = 300$

Eye Color	Red	Black	Grey	
Frequency	150	150	300	Total = 600

Because 150, 150, and 300 are each greater than 5, we can proceed.

4 Compute X^2

$$X^2 = \frac{133^2}{150} + \frac{155^2}{150} + \frac{312^2}{300} - 600$$

$$= \frac{17{,}689}{150} + \frac{24{,}025}{150} + \frac{97{,}344}{300} - 600$$
$$= 117.9 + 160.2 + 324.5 - 600$$
$$= 602.6 - 600 = 2.6$$

5 *Tabled value*

We choose $\alpha = .01$ for this illustration. Since there are three columns in the table, df $= 3 - 1 = 2$.

$X_c^2 = 9.21$

6 *Formulate a decision rule*

We will reject the null hypothesis if our outcome is greater than 9.21.

7 *Conclusion*

Since 2.6 is not greater than 9.21, we fail to reject the null hypothesis. We do not have sufficient statistical evidence to prove that the null hypothesis is wrong. That is, the genetic theory may well be true for this population.

CLASS EXERCISE

4-8 A chi-square experiment is carried out to test the hypothesis H_0: This die is fair, against the hypothesis H_a: this die is not fair. The observed value of X^2 is 45.7. If $\alpha = .01$, would the correct decision be to reject or not to reject H_0? Does the evidence indicate that the die is not fair? ____________

4-9 A chi-square experiment is designed to test the hypothesis that in a certain population $\frac{2}{6}$ of the women are single, $\frac{2}{6}$ married, $\frac{1}{6}$ widowed, and $\frac{1}{6}$ divorced. The null hypothesis is H_0: These proportions are correct. The alternative hypothesis is H_a: these proportions are not correct. If $\alpha = .05$ and X^2 turns out to be 7.42, would the correct decision be to reject or not to reject H_0? Does this indicate that the proportions given by the hypothesis are not correct?

____

4-10 A statistician conducts a hypothesis test. Using $\alpha = .01$, he concludes that the H_0 is wrong. Answer the following true or false.

a There is only a 1 percent chance that he is wrong due to the luck of sampling. ____________

b There is only a 1 percent chance that he is wrong due to not having a truly random sample. ______

c There is only a 1 percent chance that he is wrong. ______

4-11 On another test he used the .05 significance level. This time he concluded that he should not reject H_0.
True or false? He is 95 percent sure that he is correct *if* his sample was random. ______

Information for Exercises 4-12 to 4-15. A die is rolled 100 times and the number of times the result is even is noted. Here are four possible resulting situations.

A The die is fair, and the results (51 evens) lead you to conclude that it is fair.
B The die is biased, but the results (49 evens) lead you to conclude that it is fair.
C The die is fair, but the results (75 evens) lead you to conclude that it is biased.
D The die is biased, and the results (75 evens) lead you to conclude that it is biased.

4-12 The alpha refers to which situation? ______

4-13 If $\alpha = .05$, then the probability 95 percent is associated with which situation?
______

4-14 The significance level tells us nothing about cases ______ and ______ occurring.

4-15 The hypothesis that the average height of MP's is 6 feet 3 inches is tested. State the four possible situations that could result.
______
______
______
______

4-16 With which of the four outcomes do we associate α? ______

A SECOND TYPE OF CHI-SQUARE PROBLEM: A TEST FOR STATISTICAL INDEPENDENCE

Dr. I. DuSurvai the noted French sociologist was studying attitudes of young adults about divorce. Dr. DuSurvai went down to the city park in Paris one day and asked this question of young adults stopped at random: "Should divorce be *automatic* if one of the marriage partners wants it?" She wanted to see if the answer to the question depended on the sex of the person answering.

She reasoned like this. Suppose the attitude of the person does *not* depend on sex. Then men and women are equally likely to respond "yes" to the question about divorce. This means that whatever percentage of women respond "yes," she would expect the *same* percentage of men to answer "yes." And the same would hold for the percentage answering "no" or "undecided."

On the other hand, suppose the responses *do* depend on the sex of the answerer. Then the men and women should end up with *different* percentages in each type of answer. Table 4-8 shows the results of her experiment.

TABLE 4-8 OBSERVED RESULTS OF ATTITUDE TOWARD AUTOMATIC DIVORCE

Sex	For	Against	Undecided	Total
Male	114	32	14	160
Female	110	43	47	200
Total				360

By looking at these figures, Dr. DuSurvai decided that men and women do tend to answer *differently* and that, therefore, attitude toward divorce *is* dependent on sex (at least in the population she sampled).

This type of experiment examines two variables and asks whether they are statistically independent or dependent. In this example the two variables are *sex* (for which there are two categories) and *attitude toward automatic divorce* (for which there are three categories). A variable is just some characteristic of a population which is divided into categories. The table which results from putting one variable across the top and the other down the side is called a **contingency table.**

The basic idea of a contingency table is this. We can compute a set of expected values which should occur if the two variables are independent. Then we see if the observed results are close to the expected ones. If they are not, we have evidence of dependence. Since Dr. DuSurvai's experiment is another example of an investigator comparing observed and expected results, you might guess that we can also compute a chi-square statistic for the experiment. We outline the procedure to show how the calculations are used to support her intuitive interpretation of the results. Since the assumption of independence is the one which will enable us to compute the expected values, it is the null hypothesis.

PROCEDURE FOR CHI-SQUARE TEST FOR STATISTICAL INDEPENDENCE

1 Identify the population, sample, and state the hypothesis.
2 Record observed values.
3 Compute expected values. Check that they are each greater than 5.
4 Compute X^2.
5 Find X_c^2 in the table.
6 Write a decision rule.
7 State the conclusion that follows from applying this rule to your outcome.

We'll go through this procedure for Dr. DuSurvai's experiment.

1 *Hypotheses*

Sex and attitude about automatic divorce are *independent* in this population (the young adults of Paris). That is, a person's attitude does not depend on his sex. Or put another way, men and women do not differ in their attitude toward divorce. Formally, we write:
Null Hypothesis: H_0 : Sex and attitude are independent.
Alternative Hypothesis: H_a : Sex and attitude are not independent.

2 *Observed values*

On the basis of interviews with a random sample of 360 young adults, we record the results shown in Table 4-9.

TABLE 4-9 OBSERVED RESULTS OF ATTITUDE TOWARD AUTOMATIC DIVORCE

Sex	For	Against	Undecided	Total
Male	114	32	14	160
Female	110	43	47	200
Total	224	75	61	360

These 360 people are a sample of the whole population of young adults.

3 *Expected values*

We wish to fill in the six cells of Table 4-10 with numbers predicted by our null hypothesis.

TABLE 4-10

Sex	Yes	No	Undecided
Male			
Female			

How to do it?

a Enter same totals as observed data in Table 4-11.

TABLE 4-11

Sex	Yes	No	Undecided	Total
Male				160
Female				200
Column Totals	224	75	61	360

b On the basis of our assumption (the null hypothesis) that sex and attitudes are independent and since 160/360 of the sample was male, we expect 160/360 of the 224 who said yes to be male also. Therefore, to find the entry for the first cell, we multiply 160/360 × 224. Analogously, to find the expected value for any cell, multiply the row total by the column total and divide by n, that is

$$E = \frac{(RT)(CT)}{n} = \frac{160 \times 224}{360} = 99.6$$

See Table 4-12.

TABLE 4-12

Sex	Yes	No	Undecided	Total
Male	$\frac{(160)(224)}{360} = 99.6$	$\frac{(160)(75)}{360} = 33.3$		160
Female				200
Column Totals	224	75	61	360

Notice: In this table, as soon as we compute *two* entries by this technique, there is only *one* space left in each row and in each column, and therefore, the entries in these spaces are automatically determined because the row and column *totals* must hold. We therefore say this table has only *2 degrees of freedom.*

Table 4-13 corresponds to our null hypothesis. We have the same percentage of men and women in each category. For example, the percentage of men answering yes was

TABLE 4-13

Sex	Yes	No	Undecided	Total
Male	99.6	33.3	27.1	160
Female	124.4	41.7	33.9	200
Column Totals	224	75	61	360

99.6/160 = 62 percent and the percentage of women answering yes was 124.4/200 = 62 percent. This is the theoretically correct table for a survey of 160 males and 200 females if sex and attitude are unrelated. Because the six expected values are each greater than 5, we can proceed.

4 *Compute X^2*

To compare the expected and the observed results, we use exactly the same formula as before

$$X^2 = \sum \left(\frac{O^2}{E}\right) - n$$

See Table 4-14.

TABLE 4-14

O	E	O^2	O^2/E
114	99.6	12,996	130.482
32	33.3	1,024	30.751
14	27.1	196	7.232
110	124.4	12,100	97.267
43	41.7	1,849	44.341
47	33.9	2,209	65.162
			375.235

$$X^2 = 375.235 - 360 = 15.235 = 15.24$$

5 *Find X_c^2 in table*

We select $\alpha = .01$. For 2 degrees of freedom with significance level .01, $X_c^2 = 9.21$.

6 *Decision rule*

We will reject H_0 if $X^2 > 9.21$.

7 *Conclusion*

Notice that the computed value of X^2, 15.24, is greater than the tabled value, 9.21. That is, 15.24 is so far above zero, that it is past the value which splits off the upper 1 percent of possible X^2 values. It is an extremely unlikely value to get in an experiment if the two variables are in fact independent. This means that the observed results are not close to the predicted results, or that the null hypothesis is probably wrong. Recall, the null hypothesis said there was no relation between sex and attitude. Therefore, we conclude that *for the population which our sample represents* there *is* a relation between sex and attitude toward divorce. We say a person's sex and attitude toward divorce are **statistically dependent.** (There is a .01 probability that we have come to the wrong conclusion because we happened by chance to get a very distorted sample.)

DEGREES OF FREEDOM

The concept of degrees of freedom is an important one in advanced statistical theory. We will not go into it deeply in this book. In an informal way we can say that degrees of freedom measures the number of values which enter into a computation *before* all the rest of the values are automatically fixed. For example, suppose we want to find four values which add up to 87. We can pick the first three values any way we want, but then the last one is fixed. This computation has 3 degrees of freedom. Similarly, if we want to find five numbers which have an average of 30, we have 4 degrees of freedom. In geometry, if we wish to pick the three angles of a triangle, we have 2 degrees of freedom. We can arbitrarily pick the first two angles, but the third one is forced on us since the sum of the three angles must be 180°.

In these chi-square hypothesis tests there are two types of tables. If the table has *one row* and C columns, then the degrees of freedom is $C - 1$. For example, the chart of color in Table 4-15 has 1 row and 4 columns.

TABLE 4-15

Fuscia	Vermillion	Aqua	Mauve	
				$n = 100$

Since $C = 4$, there are 3 degrees of freedom. We can freely pick any three numbers to put in the cells, but the fourth one is forced on us to make $n = 100$.

The second type of table has more than 1 row and 1 column. For example, Table 4-16 has 3 rows and 4 columns.

TABLE 4-16

Answer	Brunette	Red	Blonde	Bald	Row Totals
Yes					200
No					100
Maybe					100
Column Totals	50	50	100	200	$n = 400$

We can freely fill in the first three numbers in row number 1 (Yes) and in row number 2 (No). See Table 4-17.

TABLE 4-17

Answer	Brunette	Red	Blonde	Bald	Row Totals
Yes	20	30	50		200
No	10	20	40		100
Maybe					100
Column Totals	50	50	100	200	$n = 400$

But the remaining numbers in the last row and the last column are forced on us. Thus we see that when $R = 3$ and $C = 4$, we have

$(R - 1)(C - 1) = (3 - 1)(4 - 1) = 2 \times 3 = 6$ degrees of freedom.

CLASS EXERCISE

4-17 Find the numbers that must be placed in the six empty cells in Table 4-17.

4-18 If we roll two dice for fairness, and let X equal the sum of the two faces showing, there are 11 possible outcomes for X, 2 to 12. A contingency table would be 1 by 11. How many degrees of freedom are there? __________

Find the number of degrees of freedom in the following size contingency tables and find X_c^2 for $\alpha = .01$.

	Table Size	Degrees of Freedom	X_c^2
4-19	4 by 6	____________	___
4-20	5 by 2	____________	___
4-21	1 by 7	____________	___
4-22	2 by 8	____________	___
4-23	8 by 1	____________	___
4-24	2 by 2	____________	*

* *Note:* We have not done any 2 × 2 experiments in this book because they require a correction factor. See Assignment 37. They can also be done as two-sample binomial hypothesis tests which you will learn in Chapter 6.

4-25 If a researcher rejects the null hypothesis that cholesterol and heart attacks are independent at the .01 significance level, he can then claim that cholesterol causes heart attacks. True or false? ____________

4-26 Why? ____________

4-27 He can then claim that heart attacks cause an insatiable yearning for cholesterol. True or false? ____________

We must be careful not to confuse the mathematical claim that two things are statistically *related* with the stronger implication that one thing *causes* another. A statistician might prove that inflation and strikes are not independent of one another. It is the task of the economists and other specialists to investigate the specific implications of such a relationship. Statisticians are members of a scientific team. They need input from the other scientists to help in designing the hypothesis test. They must understand enough of the scientist's field to help formulate the correct questions and analyze the conclusions. There is a world of difference between: "smoking and lung cancer are related"; and "smoking causes lung cancer"; or "lung cancer causes smoking"; or "a lack of vitamin Q causes both lung cancer and a craving for tobacco."

CLASS EXERCISE

4-28 Is there any correlation between teachers' salaries and the number of pizza parlors operating over the last 10 years? ____________

You may be surprised to learn that the answer is yes! Both have been steadily increasing.

4-29 Therefore, if more pizza parlors operate next year, your instructor will get a raise. True or false? ______

4-30 Can you think of any reasons for the correlation between these two variables?

______

Chapter Summary

1 We have seen how to compare observed and expected values from a random experiment by using the X^2 statistic. We compare the sample value of X^2 with a critical value of X^2 listed in a chi-square table. This procedure is one for testing the null hypothesis of statistical independence for data which are collected in contingency tables.

2 We have seen that the significance level of a test measures the probability of a Type I error and that a hypothesis test is not complete unless some estimate of the probability of error due to sampling is given.

Words and Symbols You Should Know

Alpha	Independence	E
Alternative hypothesis	Null hypothesis	H_a
Cell	Observed value	H_0
Chi-square table	Row	n
Column	Row total	O
Column total	Sigma	R
Contingency table	Significance level	RT
Critical value	Type I error	X^2
Decision rule	Type II error	X_c^2
Degrees of freedom	C	α
Expected value	CT	Σ

ASSIGNMENT

There was a young lass with a flair
For tests which are known as chi square
Her observed and expected
Were confused, she rejected
A perfectly honest dice pair

1 Find the critical value of X^2 in each case:

Case	Number of Rows	Number of Columns	α	df	X_c^2
a	3	5	.01		
b	1	6	.01		
c	4	8	.05		
d	4	2	.05		
e	1	7	.01		
f	1	5	.05		
g	8	1	.05		
h	4	8	.01		

2 A spinner is divided into eight equal parts and colored as shown in the figure. To test the spinner for fairness it is spun 80 times.

a How many times should a fair spinner, colored like this one show red? blue? yellow?

b If the observed results were 10 red, 50 blue, and 20 yellow, show that X^2 is exactly zero.

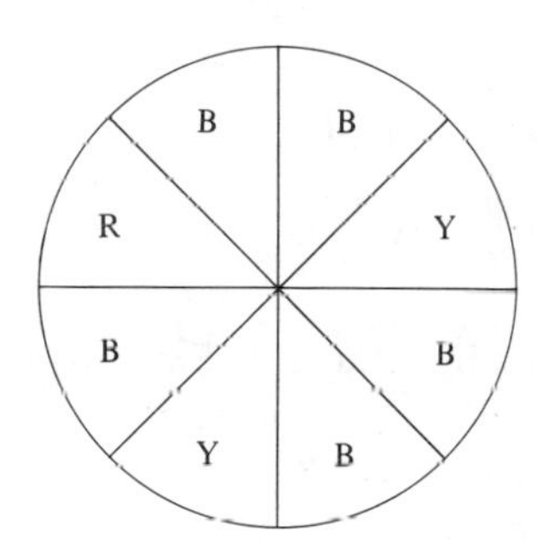

3 An investigator, May B. Knott, seeks to find out if students' grades are independent of the number of hours they work. Her analysis will be carried out at the .05 significance level. Four possible resulting situations are:

A They are independent, but the investigator rejects independence.
B They are independent, and the investigator fails to reject independence.
C They are dependent, and the investigator rejects independence.
D They are dependent, but the investigator fails to reject independence.

a In which two situations will the investigator have come to a correct conclusion?

b The probability of .05 corresponds to which situation?

c The probability of .95 corresponds to which situation?

4 A claim that a roulette wheel is fair is tested at the .01 significance level. Suppose the wheel is actually fair.

a Which situation has a probability of occurring equal to .01?

b Which one has a probability of occurring equal to .99?

5 Test a die for fairness by rolling it 90 times.

6 Can you make a fair spinner out of cardboard? Try it by constructing a spinner with four equal areas. Test it for fairness by spinning it 80 times.

7 Compute X^2 for the data in Figure 2-3. Does X^2 indicate that you should reject the null hypothesis of fairness? Let $\alpha = .05$.

8 According to a national survey in 1974, the percentages of people who say they "follow" various sports was

Football, 63 percent
Baseball, 58 percent
Basketball, 44 percent
Tennis, 26 percent

Interview 100 people on your campus (or in your neighborhood) to see if these percentages are reasonable for them.

9 A theory in biology predicts that in a large population 74 percent of the people are right handed, 20 percent are left handed, and 6 percent use both hands equally well. A survey was taken of 100 people with these results: 80 right handed, 17 left handed, and 3 use both hands.

a Draw and label a table of observed values.
b Draw and label a table of expected values.
c How many degrees of freedom are there in this experiment?
d What is the null hypothesis for this test?
e If you test with $\alpha = .05$, what is the critical value of X^2?
f What is the critical value of X^2 used for?
g Compute X^2. Does this lead you to reject the null hypothesis? Explain.

10 According to the theory proposed by the noted cultural investigator J. Hoy Poloy, 50 percent of television viewers prefer to watch sports, 30 percent prefer to watch comedy, and 20 percent prefer to watch news. A random sample of 200 viewers found 90 prefer sports, 60 prefer comedy, and the rest prefer news. Does this negate Hoy Poloy's theory? Let $\alpha = .05$.

11 In a hotly contested election for mayor of a big city it was thought that the five big ethnic groups tended to vote as blocks. A random survey of 1,400 voters from these groups was taken. Each person's ethnic group and preferred candidate was recorded. The results are listed below. The null hypothesis would be that there is no relation between ethnic group and candidate preference. Do the observed data give us good reason to reject this hypothesis at the .01 significance level?

Candidate	Black	Spanish	Jewish	White Protestant	White Catholic
Bean	100	50	100	110	150
Barildo	400	250	100	90	50
Total	500	300	200	200	200

12 Libby Raishen, a college student, did a survey of female shoppers in a neighborhood shopping center. She was investigating whether a woman's attitude toward abortion depends on whether or not she is a mother. She asked 114 women stopped at random this question: "Should it be (easier) (harder) (no different) than it is now for a woman to get an abortion?" Here are her results:

Woman's Status	Easier	Harder	No Different	No Opinion
Mother	4	8	32	10
Non-Mother	20	18	18	4

What should Libby decide on the basis of a chi-square computation, if $\alpha = .05$?

13 A professional opinion surveyor, Manon DeStreet, did a survey of female moviegoers at a neighborhood theater to investigate the relation between marital status and attitude toward women's liberation. Here are his results:

Marital Status	For	Against	No Opinion
Married	15	24	13
Single	27	18	3

What is the conclusion on the basis of a chi-square computation if $\alpha = .01$?

14 A psychological clinic administered three medicines to help relieve fear of statistics exams. It was then recorded which medicine helped each patient the most. The table shows the results of testing these medicines on 300 people. Does there seem to be a relation between which medicines work best and the person's age? Check your intuition by computing X^2. Use $\alpha = .05$.

Age	Prescription A	Prescription B	Prescription C
15–29 years	20	66	14
30–49 years	15	75	10
50–69 years	25	58	17

15 Dr. Louis L. Lewis of Nassau Hospital stated that there was a definite pattern to Emergency Room medical cases. "In the early morning you have industrial accidents. After lunch it's stomach cramps. Then some kids with broken arms, bike and skate accidents. These are followed by commuters who hurt themselves running for the train, falls (in the bar car), heart attacks, etc. Finally, you end the day with home accidents, dishwashing burns and cuts and the like." To test whether this comment is accurate or not, we surveyed the records of Nassau Hospital's Emergency Room and found the following data for the last 50 cases of these *five types* of accidents.

Type of Case	Before Noon	Noon to 3 P.M.	3 to 5 P.M.	5 to 7 P.M.	After 7 P.M.
Industrial Accidents	20	10	16	4	0
Stomach Ailments	4	16	9	14	7
Children's Accidents	9	15	15	7	4
Commuters' Complaints	9	7	5	19	10
Home Accidents	5	12	11	8	14

Do a chi-square hypothesis test on these data using $\alpha = .01$. What is the null hypothesis?

16 A nuclear plant is planned for the town of Ellenville. A researcher gathers the following data:

Education Level	Favor Nuclear Plant	Favor a Fuel Oil Plant	No Opinion
College Graduate	40	30	30
Some College	70	20	10
High School Graduate	30	50	20
Other	30	40	40

Is this data evidence that the level of education and attitude toward the nuclear plant are dependent at the .05 significance level?

17 Is one's college class independent of a choice of a career at State University? A senior, Sally Forth, gathered the following data:

College Class	Made a Career Decision	Did Not Make a Career Decision
Freshman	22	8
Sophomore	36	11
Junior	8	15
Senior	16	14

Perform a hypothesis test with $\alpha = .01$.

18 Does parents' knowledge of a foreign language affect their children's choices? Students of French were asked to check one of the following:

A Neither parent speaks a foreign language.

B Neither parent speaks French, but at least one knows some other language.

C One parent speaks French.
D Both parents speak French.

Similar questions were put to students taking Spanish and Italian. The results were as follows:

Parents' Knowledge of a Given Foreign Language	Language Student Is Taking		
	French	Spanish	Italian
No Foreign Language Known	120	110	170
Different Language Known	50	20	30
One Parent Knows Language	130	70	200
Two Parents Know Language	100	300	200

Test with $\alpha = .01$.

19 A survey of JFK High School senior girls showed the following data concerning height and wearing heels:

Height	Seldom	Sometimes	Very Often
Over 5 feet 7 inches	10	11	4
Between 5 feet 2 inches and 5 feet 7 inches	14	26	14
Under 5 feet 2 inches	5	7	9

Does this establish a relationship between heights and wearing heels? Use $\alpha = .05$.

20 Is there a relationship between smoking and coffee drinking among the customers at Sally-Ann's Truck Stop Diner? A waitress, Althea Tamara Knight, collects the following data by observing customers. Test with $\alpha = .05$.

Coffee-Drinking Category	Cigarette	Cigar	Pipe	Neither
Drank Coffee	44	21	6	17
Did Not Drink Coffee	32	13	14	23

21 Are sex and major field related at Podark Community College if a random sample of students produces the following data? Use $\alpha = .01$.

Sex	Education	Social Science	Humanities	Business-Secretarial	Natural Science	Allied Health	Other
Male	4	10	2	16	9	4	7
Female	13	11	20	7	5	13	10

22 The Ripoff Company manufactures paper clothing. They make four styles of men's shirts which sell for four different prices. They also classify their customers in three age groups: teens, twenties, over thirty. They wish to see if their sales figures indicate whether all styles are equally popular in each age group. Here are the sales figures for one month. Analyze the data at $\alpha = .05$.

	Price			
Age Group	**$1**	**$2**	**$3**	**$4**
Teens	20	80	150	150
Twenties	10	60	150	80
Over Thirty	90	80	20	10

23 Bermuda Schwartz, the master engineer at the Robot Maternity Hospital is keeping track of the eyes of baby robots to see if there is any relationship to the eyes of the mother robot. Here are the results so far:

	Mother's Eyes		
Babies' Eyes	**Steel**	**Glass**	**Plastic**
Steel	50	25	25
Glass	25	50	25
Plastic	25	25	50

Analyze these data at $\alpha = .01$.

24 A geneticist was studying saliva samples taken from three racial groups. Three "types" of saliva were discovered in each racial group. Here are the results of an analysis on random samples taken from the three groups:

Saliva Type	**Caucasian**	**Black**	**Chinese**
A	66	52	28
B	44	22	11
C	10	5	1

Does this indicate some dependence between racial background and saliva type? $\alpha = .05$.

25 Connie Kwickbill, the noted horse race expert, was testing a theory to see if there was any relation between starting position and finishing order. She kept track of the first four starting positions and the first three finishing horses.

	Start			
Finish	1	2	3	4
1	40	40	45	50
2	50	45	48	52
3	50	45	37	48
Total	140	130	130	150

On the basis of these figures, what should her conclusion be? $\alpha = .05$.

26 Susan interviewed 200 persons in another state last weekend. She observed the data in the following 5×3 chart.

Marital Status	For	Against	No Opinion
Married	40	10	0
Single	30	50	0
Widowed	15	4	1
Divorced	10	10	0
Separated	20	7	3

It is clear that she cannot get all 15 expected values greater than 5 as was suggested earlier in this chapter. What changes would you advise her to make in order to garner some results from her experiment?

27 "Buck" Ed Sietz counted 100 cars on campus. His data is summarized below:

Make of Car	Sedans	Con-vertibles
GM	50	10
Ford	16	4
Chrysler	14	6

Two of his expected values are not greater than 5. What should he do to salvage this experiment?

28 An experiment was done to see if honeybees communicate the direction to food just by their "dance" when they return to the hive. Another idea was that they depended on the scent of the food. The following experiment was set up as shown in Figure A4-28. The empty squares represent empty plates which were given the same scent as the food. The food plate was placed as shown by the shaded square. At the start of the experiment some bees were placed on the food plate. They returned to the hive, and immediately bees started flying out of the hive looking for the food. As they reached the various plates they were counted. The number arriving at each plate is shown on the diagram. Test the

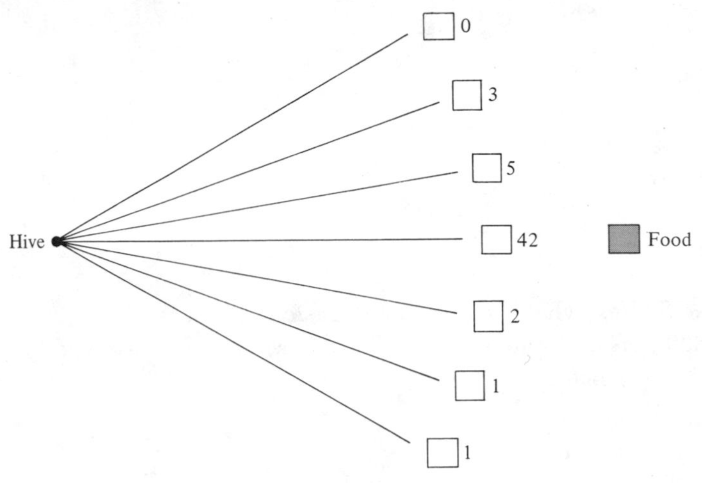

Figure A4-28

null hypothesis that all the plates are equally likely to be visited. Use $\alpha = .01$. What does this imply for the theory that the bees use scent to fly to the food?

29 a Outline a chi-square hypothesis test that you could perform. State clearly the issue being investigated, the population being tested, the method of obtaining a random sample. What *exactly* will you ask each person interviewed? What are the hypotheses? Which significance level do you intend to use? How large will the sample be? Label a contingency table with the cells that you intend to use.

b After your instructor has approved your outline, gather the data and perform the hypothesis test. Be sure to include in your report any changes you were forced to make in the original plan. Comment on the strengths and weaknesses of your conclusions.

Here is an example of how one student answered Assignment 29a. "I want to see if there is any relationship between marital status and the attitude of women toward premarital sex in Centerville. I will survey 150 women in the shopping center on West Street on Saturday morning starting at 10 A.M. I will say to each woman interviewed:

1 Pardon me, I am doing a survey for my college statistics class. Would you mind answering a few questions?
2 Do you live in Centerville?
3 Are you single, married, divorced, or widowed?
4 Do you think that premarital sex is: always wrong, sometimes wrong, or never wrong?
5 Thanks for your cooperation.

I will interview women in the parking lot. I will walk through the parking lot, up and down every row asking each woman I meet. If several women are together, I will ask only the one nearest me. If a woman refuses to answer, I will not count her. I will continue until I get 150 responses. I will record my data as follows:

Marital Status	Always Wrong	Sometimes Wrong	Never Wrong
Single			
Married			
Divorced			
Widowed			

I will only be asking women in the shopping center. A lot of women in Centerville will be excluded. However, I do not think that this will bias my results since I doubt that there is any connection between shopping and my questions. Saturday morning is a good time to get a cross section of many shoppers."

30 Make up your own question and use the data in Appendix A to answer it. For example, are age and religion independent in this population?

31 Read Chapter 8 in D. Huff's book "How to Take a Chance" (Norton, New York, 1959) for an explanation of significance level and hypothesis testing.

Readings from J. Tanur (ed.), Statistics: A Guide to the Unknown," Holden-Day, San Francisco, 1972.

32 Does Inheritance Matter in Disease? The Use of Twin Studies in Medical Research, Reid, p. 77. If inheritance does not matter, then we expect that the proportion of identical twins who both get the disease will be the same as the proportion of nonidentical twins who both get the disease. What are the results for various diseases?

33 Death Day and Birthday, Phillips, p. 52. Are they independent?

34 Measuring Racial Integration Potentials, Berry, p. 326. If Chicago would achieve a "colorblind" housing market, how could the expected proportion of black families in each neighborhood be computed? What is the "redistributive shift"?

35 Statistics, The Sun, and The Stars, Whitney, p. 385. What does "expected and observed" have to do with double stars?

Readings from The New York Times

36 Was the first Draft Lottery Random? Sunday, January 4, 1970, p. 66. Can you spot the value of alpha?

37 Remember that the chi-square table in the appendix is constructed by knowing which results for an experiment are common when the experiment is repeated many times. It has been shown that the table is not as dependable for problems which have 2×2 contingency tables as it is for larger tables, especially when n is small. To improve the dependability of the table for a 2×2 case, we use what is called a "correction factor" in the calculation of X^2. First, you compute the four expected values. Then look at the expected values and the corresponding observed values. You will notice that two of the observed values are larger than the corresponding expected values. The correction factor is simply this: subtract .5 from *these two* observed values and add .5 to the other two. Then using these adjusted, observed values proceed in the usual manner. Using this approach solve the following problem. In a study for her doctoral thesis in animal behavior, Annie Mahl gathered this data: 28 out of 70 left-handed spider monkeys preferred mangos to guavas, while 19 out of 30 right-handed spider monkeys preferred mangos to guavas. Should we conclude at the .05 significance level that there is a difference in fruit preference between left- and right-handed monkeys?

FIVE

Testing Proportions—Binomial Experiments

To be, or not to be: That is the question.

Shakespeare, Hamlet

"It's long," said the Knight, "but it's very, very beautiful. Everybody that hears me sing it—either it brings tears into their eyes, or else—"
"Or else what?" said Alice, . . .
"Or else it doesn't, you know."

Lewis Carroll, Through the Looking Glass

STATISTICAL SYMBOLS

In any course in statistics it is necessary to use symbols to represent various results. Sometimes we get a result by doing an experiment with a sample from a population; sometimes we get a result because of some theory we have about a population. It is the usual practice to use Greek letters as symbols for theoretical results and to use English letters for experimental results. For example, if we compute the percentage of females in a sample from some population, we would use the letter p to represent our result (since p is the first letter of "*p*ercent"). On the other hand, if we were told that *according to some theory,* 80 percent of all people are right handed, then we would use the Greek for p, π (pi), to represent that result.*

Greek letters often represent results which are assumed to be *true for an entire population* because such a result is usually obtained from some theoretical discussion and not from some sampling experiment. A sampling experiment cannot usually cover an entire population. A value which is true for an *entire population,* such as the true mean of a population (say, the average age of all American citizens next Saturday at 1:00 P.M.), is called a **parameter.** A value based on a sample from a population is called a **statistic.** For example, in Chapter 4 the X^2's that we computed are statistics. From what we said earlier, Greek letters usually represent parameters and English letters usually represent statistics.

Here is a brief list of some pairs of English and Greek letters and what they are used for in this book. The individual pairs are explained whenever we get to them in the text.

TABLE 5-1

English	Greek	What It Represents
p	π (pi)	*p*ercent or *p*roportion or *p*robability
m	μ (mu)	*m*ean (average)
s	σ (sigma)	*s*tandard deviation
r	ρ (rho)	cor*r*elation coefficient

CLASS EXERCISE

5-1 Suppose you get a sample of 15 protozoa and find that the *m*ean (average) time it takes for one to split in half is 17.4 hours. You would write ______ = 17.4.

* *Reminder:* This use of the symbol π does not stand for 3.14 · · · but rather for the theoretical value of a *p*robability. See Table 5-1.

5-2 Suppose you read in a biology book that according to current cell theory the *mean* time it takes a certain type of protozoa to split in half is 22 hours. You write _____ = 22.

INTRODUCTION TO THE NORMAL CURVE

In Chapter 4, we saw that by repeating certain experiments many times we could come to know which results were common and which were rare. We saw that statisticians can describe the pattern of results which should occur in certain repeated experiments. In particular we saw that the pattern of X^2's is described by a curve called a **chi-square curve.** Knowing this curve allows us to make tables of critical values of X^2 for use in other experiments.

In this chapter, we are going to be computing a different statistic which we shall represent by the symbol p. We shall see that statisticians can describe the pattern of p's which results from repeated sampling. This pattern is given by a curve called a **normal curve.** It is worth discussing the normal curve by itself before we get to the experiments in which we compute p.

An example of a normal curve is shown in Figure 5-1. When a normal curve is used to describe the results of some experiment, it is always shown with a horizontal axis below it. Here is an example (Figure 5-2) of a normal curve used to represent a distribution of weights. A **distribution** is simply a *collection* of numbers. We interpret Figure 5-2 as follows. Because the graph is highest above the number 150, we know that weights near 150 pounds are common in this population. Because the graph gets lower as we go away from 150, we know that weights far from 150 are rare in this population.

If we shade in two parts of the graph as shown in Figure 5-3,

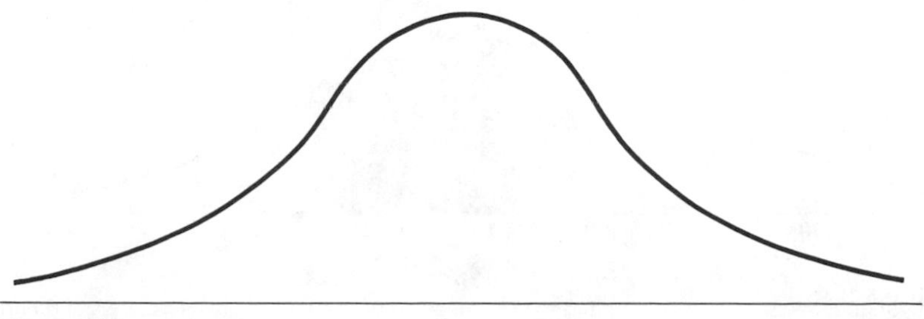

Figure 5-1

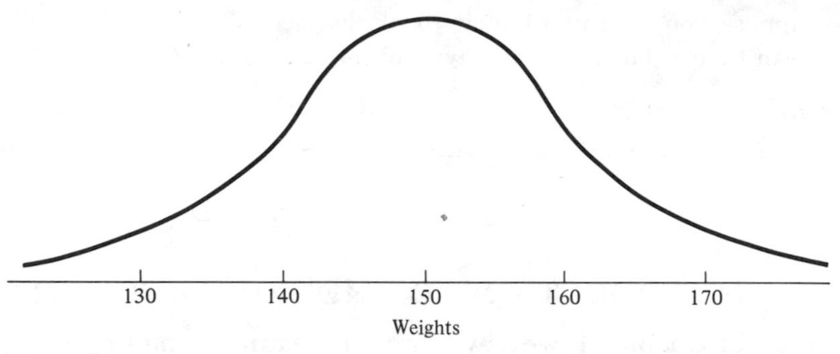

Figure 5-2

we can see that there is much more area in section A than there is in Section B. This means that there are many more weights between 150 and 160 pounds than there are between 160 and 170 pounds in this population.

When the graph for a distribution of measurements is a normal curve, we say that the measurements are **normally distributed.** Since a normal curve is symmetrical, the number on the axis directly beneath the highest point on the normal curve is the **mean** of the distribution. It is the average of all the measurements.

CLASS EXERCISE

5-3 The normal curve in Figure 5-3 describes a distribution of weights in which the average weight is ____________________ .

5-4 For the population of weights described in Figure 5-3, there are a certain number of weights between 130 and 140 pounds. There are about the *same* number of weights between ________ and ________.

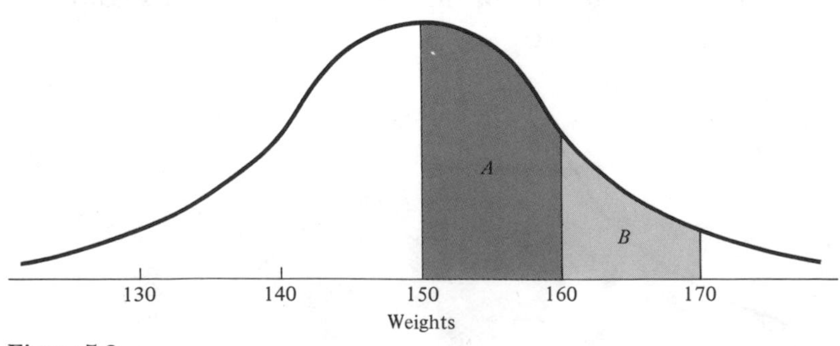

Figure 5-3

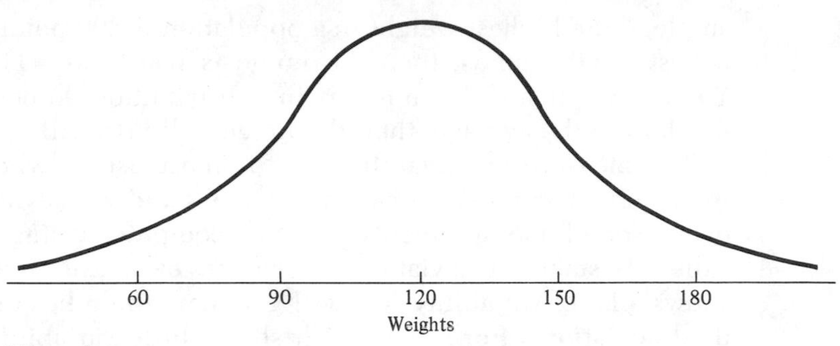

Figure 5-4

Measuring Variability

Look at the two normal curves shown in Figures 5-4 and 5-5. In Figure 5-4 the average weight for the distribution is 120 and weights between 60 and 90 appear often enough to occupy a substantial part of the area under the curve. In Figure 5-5 the average is also 120, but weights between 60 and 90 are so rare that they occupy practically no part of the area under the curve.

We summarize this by saying that the weights in Figure 5-4 show *more variability* than the weights in Figure 5-5. In Figure 5-5 the average weight is 120 and almost all the weights are between 100 and 140. In Figure 5-4 we see that there are plenty of weights less than 100 and more than 140. In Figure 5-5 most of the weights are closer to the average; they do not vary very much from the average.

Statisticians have several ways to measure variability. The general idea is that any way you decide to measure variability should give a large number to a set of measurements which spread out far on both sides of the mean, and a smaller number to a set which does not spread out as much. The simplest measure of variability is the **range**. The range of a set of measurements is the difference between the highest and the lowest values. For ex-

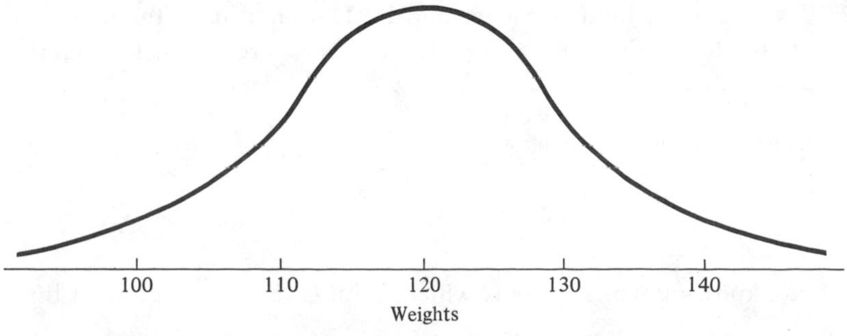

Figure 5-5

ample, if the highest weight in a population is 200 pounds and the lowest is 90 pounds, then the range is $200 - 90 = 110$ pounds. You can see that if the numbers in a distribution do not vary very much from the average, then the range will be small.

The measure of variability that is most useful when dealing with normal curves is one called the **standard deviation.** For every set of measurements we can compute a single number called its standard deviation. For a list of measurements which shows a lot of variability, we will get a large number for the standard deviation. For a list which shows little variability, we will get a small number for the standard deviation. If we have arrived at the value of the standard deviation by some theory, we use the Greek letter sigma, σ, as the symbol to represent it. If we have used sample data to arrive at it, we use s for the symbol. We will explain later various ways to compute the standard deviation. For now, we just say that if one distribution has more variability than another, then it will have a larger standard deviation. For example, the normal distribution of weights in Figure 5-4 has a larger standard deviation than the one in Figure 5-5. It might be, perhaps, that the standard deviation in Figure 5-4 is 30 pounds and in Figure 5-5 it is 10 pounds.

CLASS EXERCISE

5-5 The grandchildren in two large families have an average age of 26 years. The first family has a range of 14 years and a standard deviation of 4 years. The second family has a range of 10 years and a standard deviation of 3 years.

Which family is likely to have more grandchildren under age 21? ____________

5-6 Fizz soft drink machines dispense on the average 8 ounces of soda per cup with a standard deviation of $\frac{1}{4}$ ounce. Kwench machines also dispense an average of 8 ounces, but the standard deviation is $\frac{3}{4}$ ounce. Which machine is more likely to spill soda out of a 9-ounce cup? ____________

5-7 Two students, John Q. Smarts and Ira Dummer, have the same biology teacher. On the teacher's desk they see the results of a recent test which they both took.

Class	Class Average	Standard Deviation
1	78	4
2	77	7

Not knowing which class is which, John Q. Smarts hopes that his is class 2, but his friend Ira prefers to be in class 1. Why? ____________

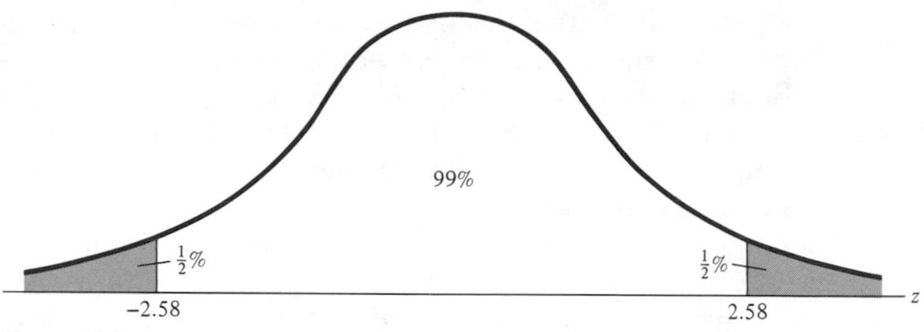

Figure 5-6

Using the Standard Deviation

Often we want to say whether our experiment has produced a usual or an unusual outcome. One way to do this is to say how far the outcome was from the average outcome. A statistician often does this in terms of standard deviation by using statements like, "This outcome is 3 standard deviations above the average." For example, in a distribution of weights with mean 150 pounds and standard deviation equal to 10 pounds, a weight of 180 pounds is 3 standard deviations above the mean.

Many facts about normally distributed sets of values are given in terms of standard deviations. For example, it can be proven that if a set of values is normally distributed, then 95 percent of those values will fall within 1.96 standard deviations of the mean, and 99 percent of those values will fall within 2.58 standard deviations of the mean.

We will often illustrate these facts by labeling sketches like the two in Figures 5-6 and 5-7. The numbers 1.96 and 2.58 are known as **z scores.** A z score is *a number of standard deviations away from the mean.* A z score of +1.96 denotes 1.96 standard deviations *above* the mean. A z score of −2.58 denotes 2.58 standard deviations *below* the mean. A z score of 0 denotes a score which is exactly equal to the mean.

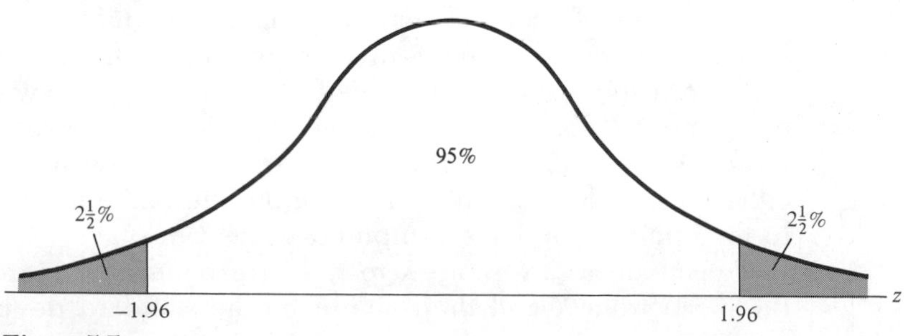

Figure 5-7

To illustrate the idea of z scores, suppose that a certain distribution of weights has a mean equal to 120 pounds and a standard deviation equal to 30 pounds. Then a z score of +2 would correspond to a weight which was 2 standard deviations above 120 pounds. That would be 60 pounds above 120 pounds or 180 pounds. Similarly, a z score of -1 corresponds to a weight 30 pounds *below* average. That would be 90 pounds.

Suppose you are working on a problem with a distribution of weights and you have learned that

1 The distribution is normal in shape.
2 The mean is 100 pounds.
3 The standard deviation is 12 pounds.

One way to record this information is like this:

$$nd_W \ (\mu = 100, \sigma = 12)$$

In this notation:

nd represents *n*ormal *d*istribution
W represents *w*eights
μ represents mean (μ, "mu," is a Greek *m*)
σ represents standard deviation (σ, "sigma," is a Greek *s*)

Similarly, if you read in this book the expression nd_X ($\mu = 50$, $\sigma = 4$), you know that some set of measurements, which we have called X, is normal with mean 50 and standard deviation 4. When you give the mean and standard deviation of a normal distribution, you have described it completely. We shall see later that knowing these two things allows us to figure out anything else we might want to know about the distribution. For example, if we have nd_W ($\mu = 120$, $\sigma = 30$), then we can answer questions like, "What percentage of the weights in this distribution are more than 150 pounds?"

Formulas for z scores

We often need to do computations with z scores. For example, we may want to know what weight corresponds to a certain z score. Or we may have the inverse problem of finding out what z score corresponds to a certain weight. In general, we will have an experimental observation of some type and its corresponding z score. We call the experimental observations **raw scores.** If you collect a sample of people and compute the percentage of females in the sample, you have computed a raw score.

To convert a raw score X to its corresponding z score subtract the mean from X and then divide by the standard deviation.

In symbols
$$z_X = \frac{X - \mu}{\sigma}$$

To convert a z score to a raw score, multiply the z score by the standard deviation and add this product to the mean. In symbols
$$X = \mu + z\sigma$$

CLASS EXERCISE

5-8 It is well known by one or two people that IQ scores of nematodes are normally distributed with mean 500 and standard deviation 50. Figure CE5-8 is a graph of these scores.

a Complete the label at the top of the graph.

b What number should be at the center of the IQ axis? __________

c What number should be at the center of z-score axis? __________

d Why is the graph highest in the middle? __________

What does this tell us about nematodes? __________

e What would be the IQ of a nematode if its z score were -1? __________

f A nematode is considered to be a "brain" if its IQ is over 598. Melba the nematode has a z score of 2. Is Melba a "brain"? __________

g What percentage of the nematodes are "brains"? __________

5-9 Suppose a normal distribution of numbers which we have called p has mean equal to .70 and standard deviation equal to .04.

a This symbolized by nd_p (_____ = _____, _____ = _____)

b What value of p has a z score of $+1$? __________

c -1? __________

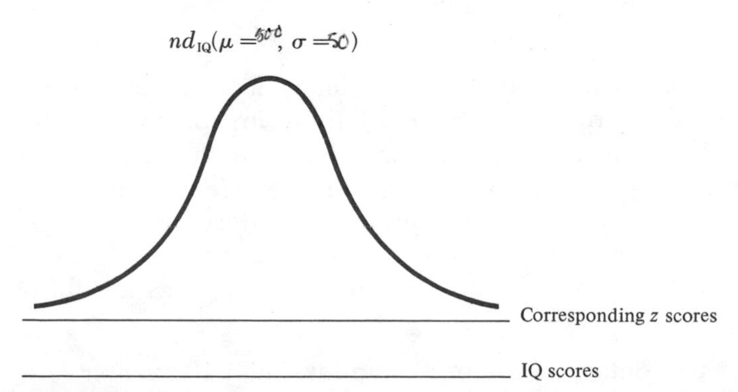

Figure CE 5-8

✱ d +2? ______________________

e +1.96? ______________________

f −1.96? ______________________

g +2.58? ______________________

h −2.58? ______________________

5-10 A sample of Martians have an average height of 21 feet and a standard deviation of 2 feet.

a Peggy has a height of 25 feet. What is his z score? ______________

b Joseph has a z score of −1. What is her height? ______________

c Patricia is 21 feet tall. What is his z score? ______________

d Complete the following chart.

Height, Feet

Martian	Sex	Raw Score	z Score
Ramon	F	19	
Victor			4
Gloria			−3
Gladys		22	
Vincent			0

Using the Normal Curve

Let Y represent a distribution of ages of pine trees in a particular forest. The symbol nd_Y ($\mu = 110$, $\sigma = 8$) says three things about this distribution of ages:

1 The distribution is a normal distribution.
2 The average age is 110 years.
3 The standard deviation is 8 years.

Mathematicians have computed many facts about the graph of the normal distribution. Some of them are listed in Table 5-2. Although the mean and standard deviation of normal distributions may vary widely, the relationship between z scores and areas does not change. You can convert the information in this table just exactly the way you usually convert z scores and raw scores. Using this table, we can answer such questions as the following. Given: nd_Y ($\mu = 110$, $\sigma = 8$),

1 What percentage of the trees live less than 102 years?
2 2.3 percent of the trees live more than what age?

3 The *middle* 68 percent of the population have lifetimes between what two ages?

TABLE 5-2 TABLE OF AREAS UNDER A NORMAL CURVE TO THE *LEFT* OF z

z Score	Proportion of Area to the Left of z	z Score	Proportion of Area to the Left of z
−4	.00003	1	.84
−3	.001	1.65	.95
−2.58	.005	1.96	.975
−2.33	.01	2	.977
−2	.023	2.33	.99
−1.96	.025	2.58	.995
−1.65	.05	3	.999
−1	.16	4	.99997
0	.50		

SOLUTIONS

Hint: It is often best to sketch and label a normal curve to help visualize the problem better.

1 We change the *raw* score of 102 to a z score.

$$z_{102} = \frac{102 - 110}{8} = -1$$

In Table 5-2, 16 percent corresponds to $z = -1$, therefore, the answer is that 16 percent of the trees live less than 102 years. See Figure 5-8.

2 Since Table 5-2 gives percentages *below* a z score and not above, we subtract 2.3 percent from 100 percent and get 97.7 percent. In the table this corresponds to $z = 2$. Converting

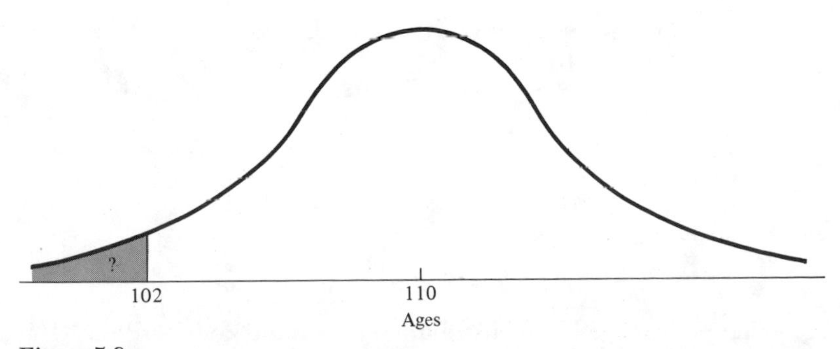

Figure 5-8

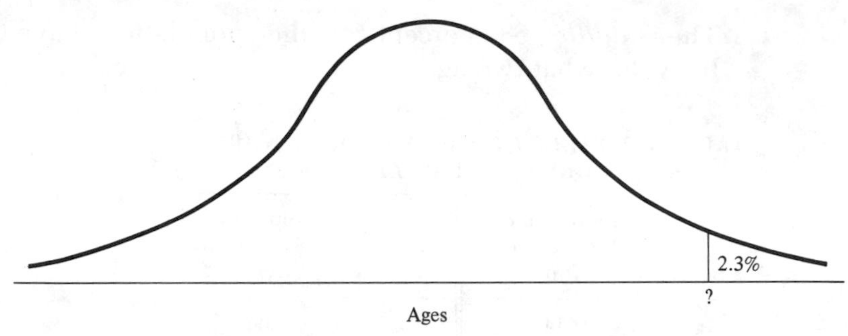

Figure 5-9

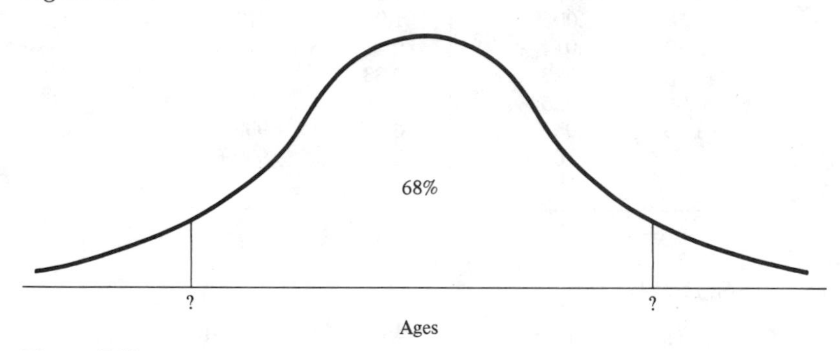

Figure 5-10

$z = 2$ to a raw score, we get

$$X = 110 + 2(8) = 110 + 16 = 126$$

Hence, 2.3 percent live more than 126 years. See Figure 5-9.

3 If the middle 68 percent lies between z_a and z_b, then the remaining 32 percent lies half to left of z_a and half to the right of z_b. Hence, there is 16 percent below z_a and $16 + 68$ percent $=$ 84 percent below z_b. Reading the table, we have 16 percent at $z_a = -1$ and 85 percent at $z_b = +1$. Therefore,

$$X = 110 + (\pm 1)\ (8) = 102 \text{ and } 118$$

or 68 percent have lifetimes between 102 and 118 years. See Figure 5-10.

CLASS EXERCISE

Given a distribution of weights.

$$nd_X\ (\mu = 50 \text{ pounds}, \sigma = 4 \text{ pounds})$$

5-11 What percentage of the distribution is under 58 pounds? ____________

5-12 What percentage of the distribution is over 54 pounds? ____________

5-13 What percentage of the distribution is between 38 and 54 pounds?

____

5-14 What weight cuts off the bottom $2\frac{1}{2}$ percent of the distribution? ________

5-15 What weight cuts off the top 16 percent of the distribution? ________

5-16 What weights cut off the middle 99 percent of the distribution? ________

5-17 *True or False* Given a distribution of people's ages with $\mu = 12$ years and $\sigma = 2$ years. 99.9 percent of the people are over 6 years of age. ________

5-18 The distribution of weights of all adults in the United States is approximately normal. Would you like your z score to be about 3, 2, 1, .5, 0, −.5, −1, −2, or −3? ________

ASSIGNMENT

The probability's not high
To pick a horse to win, so I
Never yet bet
$1 to get
A fortune, that's π in the sky

1 *True or False* Just as a small range indicates little variability, so a small standard deviation indicates little variability in a list of numbers.

2 a The distribution of the number of math classes cut per semester by nonmath majors has a mean of 8 and a standard deviation of 3. *True or False?* No student cut more than 11 classes.

b The average number of hours per week that students spend in the college cafeteria is 10. The range is 12. *True or False?* No student spent more than 16 hours in the cafeteria.

3 Standard Rules Inc. buys precut lumber from two companies for its yard sticks. It finds that wood from Company A has an average length of 36 inches with a standard deviation of .01 inch, while the wood coming from Company *B* has a mean length of 36 inches with $\sigma = .02$ inch. Which company supplies better wood for yard sticks?

4 "Dr. Ima Hevy has shown at the .01 significance level that cholesterol and heart trouble are related." from *Newsmonth Magazine*
True or False This means that in 99 out of 100 heart trouble cases, cholesterol has some influence.

5 A distribution of weights of people has the following properties: nd_W ($\mu = 180$ pounds, $\sigma = 10$ pounds)

a The middle 99 percent of the people weigh more than ________ but less than ________.

b Frances weighs 173 pounds. Find z_{173}.
c George has a z score of 2.1. Find his weight.

6 A distribution of people's ages has a mean of 43 years and a standard deviation of 5 years.
a Margret is 51. Find the z score for her age.
b Frederick's age has a z score of -3.2. How old is he?
c The middle 95 percent of these people are older than ________ but younger than ________. (Be careful!)

7 In a normal distribution of ages, what percentage
a Lie within two standard deviations of the mean?
b Are above $z = 1$?
c Are between $z = 1$ and $z = 1.96$?
d Lie within 3 standard deviations of the mean?
e Are more than 2 standard deviations below the mean?
f Are more than 2 standard deviations away from the mean?

8 In a normal distribution of people's weights, $\mu = 180$ pounds and $\sigma = 10$ pounds.
a What percentage of the people weigh less than 200 pounds?
b What percentage of the people weigh more than 160 pounds?
c 95 percent of the people centered around the average weight between ________ and ________ pounds.
d 84 percent of the people weigh more than ________ pounds.
e 99.5 percent of the people weigh less than ________ pounds.

9 In a normal distribution of biology grades, the mean was 43 and the standard deviation was 3.
a What percentage of the students scored over 49?
b What percentage of the students scored over 34?
c What percentage of the students scored between 40 and 46?
d 99.5 percent of the students scored more than ________.
e 95 percent of the students are centered about the average between ________ and ________.

10 In a normal distribution of glyphs, the mean is 30 plots and the standard deviation is 4 plots.
a What percentage of glyphs are between 34 and 38 plots?
b What percentage of glyphs are between 38 and 42 plots?
c What percentage of glyphs are less than 26 plots?
d 84 percent of glyphs are more than ________ plots?
e 99 percent of glyphs are centered around the average between ________ and ________ plots.

11 In a distribution of heights $\mu = 5$ feet 10 inches and $\sigma = 3$ inches. What percentage of the people are over 6 feet 1 inch?

12 A distribution of 2,000 IQs is approximately normal. $\mu = 110$ and $\sigma = 15$. Approximately how many persons have an IQ over 95?

A BINOMIAL EXPERIMENT

Dr. Will Ketchum, a professor of English, states that 30 percent of students who turn in book reports to him have not, in fact, read the book they are reporting on. To test the accuracy of his statement, it would be impractical, if not impossible, to check every student who prepares a book report. However, if we gather a random sample of students, and if the professor is correct, we would expect about 30 percent of our sample to have submitted phoney book reports. Another way of saying this is that if 30 percent of the population submitted phoney reports, then for any single person picked at random from the population the probability is .30 that *he* submitted a phoney report.

This kind of experiment is called a **binomial experiment.** A binomial experiment is characterized by two properties.

1 There are exactly two possible outcomes to the experiment. In our problem a student has either *read* or *not read* the book reported on.
2 If π stands for the probability of one of the two outcomes occurring, then the value of π does not change during the course of the experiment. In our problem we are assuming that the proportion of students who turn in phoney book reports does not change during our investigation, and that therefore, the *probability* of picking such a person does not change.

CLASS EXERCISE

5-19 Suppose that you get a random sample of 200 students from the professor's population and only 28 percent of the students admit to turning in a false book report. Would that prove that the professor was wrong in claiming 30 percent?

____

5-20 Would it prove that he was correct? ______________________________

5-21 What if your experimental outcome was 4 percent? ______________________________

If we assume that the 30 percent figure is accurate for the population being investigated, then we expect that a random sample from that population will also produce *about* 30 percent of the students with book reports on unread books. The question that we are asking is: What does the word "about" mean? Is 28 percent near to 30 percent? Is 40 percent? Is 15 percent? Where do we draw the line?

It can be shown (and in this chapter it will be shown) that:

1 If the 30 percent figure is correct, and
2 If you select many random samples of 200 students each, then 99 percent of the time you will get between 22 and 38 percent of the students in your samples reporting on unread books.

Thus, if you took only one sample of 200 students and the sample outcome was 28 percent, that would *not* be unusual and the 30 percent figure claimed by the professor might well be correct. However, if the sample outcome was 40 percent, it would be very unlikely that the 30 percent figure is correct. It could be, but it is highly improbable. The statistician would reject the assumption that 30 percent is correct and state that it is probably incorrect.

COMPUTER SIMULATION

We said above that if 30 percent is the true percent of cheating students, then samples of size 200 would almost always contain between 22 and 38 percent cheaters. A computer simulation can illustrate this. We programmed a computer to contain a very large population of "students." Exactly 30 percent of the students in this population had not read the book they reported on. We say that π, the true percent (which is a population parameter), equals 30 percent or .30. The computer was instructed to select a random sample of 200 "students" and to record how many had read the book reported on. Then the computer was instructed to repeat this process many times. The results of the first five computer samples are given in Table 5-3.

TABLE 5-3

Sample	Number of Students Who Had Not Read Book	Number of Students Who Had Read Book	Percentage of Students Who Had Not Read Book
1	63	137	31.5% = .315
2	64	136	32.0% = .320
3	72	128	36.0% = .360
4	58	142	29.0% = .290
5	57	143	28.5% = .285

We repeated this experiment 100 times, then we arranged the 100 results in numerical order. These appear in Table 5-4. We let π stand for the true population parameter. We use the symbol p to stand for the values in the last column since they are sample percents. The set of results in Table 5-4 is called a **distribution of sample percents or a distribution of sample**

TABLE 5-4 EXPERIMENTAL RESULTS ARRANGED IN NUMERICAL ORDER

Experiment Number	*p*	Experiment Number	*p*	Experiment Number	*p*
1	.240	35	.290	68	.320
2	.245	36	.290	69	.325
3	.250	37	.290	70	.325
4	.250	38	.290	71	.325
5	.250	39	.290	72	.325
6	.250	40	.290	73	.330
7	.255	41	.295	74	.330
8	.255	42	.295	75	.330
9	.255	43	.295	76	.330
10	.255	44	.295	77	.330
11	.260	45	.295	78	.330
12	.265	46	.295	79	.335
13	.265	47	.295	80	.335
14	.270	48	.295	81	.335
15	.270	49	.295	82	.335
16	.270	50	.295	83	.335
17	.275	51	.295	84	.340
18	.275	52	.300	85	.340
19	.275	53	.300	86	.340
20	.275	54	.305	87	.345
21	.275	55	.305	88	.350
22	.275	56	.305	89	.350
23	.280	57	.305	90	.350
24	.280	58	.310	91	.350
25	.280	59	.310	92	.355
26	.280	60	.310	93	.355
27	.280	61	.315	94	.355
28	.280	62	.315	95	.355
29	.285	63	.315	96	.355
30	.285	64	.320	97	.355
31	.290	65	.320	98	.360
32	.290	66	.320	99	.360
33	.290	67	.320	100	.365
34	.290				

Distribution of Sample *p*'s.

***p*'s.** It is easier to see the pattern of the results in a graph. To make a graph, first we group the *p*'s in convenient intervals* in a frequency table (see Table 5-5). Then we get the graph shown in Figure 5-11.

Interpretation of Computer Experiment

In theory, we would expect each value of *p* to be near 30 percent, some higher and some lower. If we averaged our 100 outcomes, we would expect the mean (or average) to be very close to

* The choice of intervals is up to the person doing the graph. A graph with 5 to 10 intervals altogether is usually convenient for showing any trends in the data.

TABLE 5-5

Intervals (Include Right End Point)	Frequency (Number of *p*'s in Each Interval)
.20 to .24	1
.24 to .28	27
.28 to .32	40
.32 to .36	31
.36 to .40	1

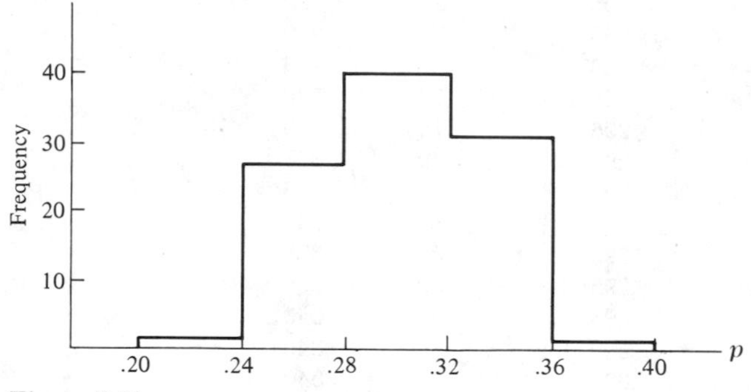

Figure 5-11

30 percent. The mean of our 100 trials was 30.4 percent.

We said previously, that on the basis of statistical theory we expected almost all the results to be between 22 and 38 percent. You can see that in this experiment actually *every* result was between 22 and 38 percent. This means that if 30 percent of the students cheat on book reports, it would not be very surprising for a survey of 200 students to turn up somewhere between 22 and 38 percent cheaters. But if the results were outside these limits, we would have strong reason to claim that the 30 percent claim was wrong.

The Normal Curve and the Binomial Experiment

A statistician reports the results of this type of experiment by giving three things:

1 The general *shape* of the graph of results.
2 The *mean*, or average, value of the results.
3 A measure of the variability of the results, usually the *standard deviation*.

What can we say about our results?

1 You can see, if you "smooth out" the graph in Figure 5-11 (as we did in Figure 5-12), the graph is bell shaped.
2 The average value seems to be around .30 which is not surprising since $\pi = .30$ was the true population parameter.

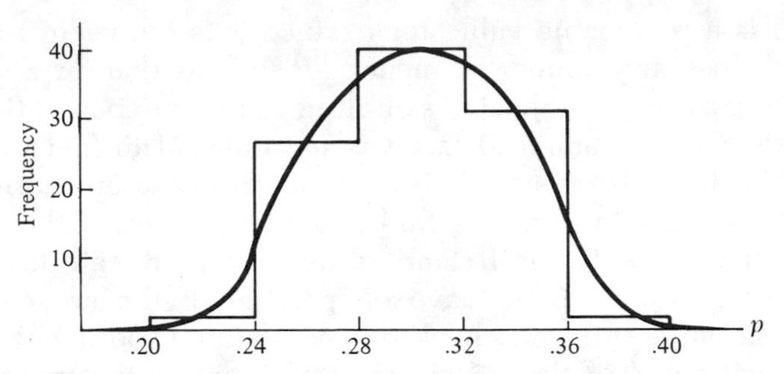

Figure 5-12

3 To compute the standard deviation, we need to introduce a new formula. Before we do that we want to talk about the general idea of a deviation so that the formula will seem reasonable.

By the word **deviation** a statistician means "distance from the mean." To find the deviation for a number, you subtract the mean from it. For the list of p's in Table 5-4 the mean is about .30. The largest p was .365. The deviation for this p, therefore, is $.365 - .30 = +.065$. The smallest p was .240. The deviation for this p is $.240 - .30 = -.060$. The negative sign simply means that the number .240 is *below* the mean. In any case, these deviations tell us that no number in the list is farther than .065 from the mean. Most of the numbers are much closer than that. Each p in the list has its own deviation.

The standard deviation σ is supposed to be a typical deviation. It is supposed to be representative of all the deviations in the list. It, therefore, should be somewhere between the largest and smallest deviations in the list. The smallest deviation in our list is 0. The deviation for result number 52 is 0 because $.300 - .30 = 0$. Therefore, we expect σ to be between 0 and .065.

The formula which statisticians give for finding the standard deviation in experiments of this type is

$$\sigma_p = \sqrt{\frac{\pi(1-\pi)}{n}}$$

where σ_p means "standard deviation of the p's," π is the true value of the parameter, and n is the sample size. In our experiment $\pi = .30$, the true percent of cheaters; and $n = 200$ since each sample consisted of 200 students. Therefore,

$$\sigma_p = \sqrt{\frac{.30(1-.30)}{200}} = \sqrt{\frac{(.30)(.70)}{200}} = \sqrt{\frac{.21}{200}}$$
$$= \sqrt{.00105} = .032$$

This is a reasonable value for σ since it is between 0 and .065. What does the number σ_p mean? It tells us that for a very long list of p's taken in repeated sampling from some population a typical deviation is about .032. It is not unusual to find values of p that far from the average. In fact, according to our normal curve table, we should expect to find about 68 percent of the p's closer than that since .032 is 1 standard deviation. Recall that between $z=+1$ and $z=-1$ we have 68 percent of the area under the normal curve. If you check our computer experiment in Table 5-4, you will find that 65 of the 100 p's are in the group ranging from .032 below .30 to .032 above .30.

Summarizing the Theory

According to statistical theory there are certain times when the sample distribution of p from a binomial experiment is approximately normal. Experiments have shown that the graph of sample values of p's will be very close to normal provided that the samples from which you get the p's are big enough. Recall that a binomial experiment is one which has two possible outcomes. We have let π stand for the probability of one of these outcomes, and, therefore, $1-\pi$ stands for the probability of the other. Experiments have shown that n, the sample size, is large enough to make the graph of p's approximately normal when the two numbers $n\pi$ and $n(1-\pi)$ are both greater than 5. In the case of the book report experiment, $\pi = P(\text{a student cheated}) = .30$ and $1-\pi = P(\text{a student did not cheat}) = .70$. Therefore, $n\pi = 200(.30) = 60$ and $n(1-\pi) = 200(.70) = 140$, and we did notice that the graph of p's seemed normal in appearance.

To describe a normal curve completely, we should know its mean μ, and its standard deviation σ. We can then use μ and σ to convert raw scores to z scores or z scores to raw scores, and thus answer any questions we might have about the measurements involved in our experiment. The following theorem states how to find μ and σ for the sample distribution of p.

THEOREM ABOUT SAMPLING DISTRIBUTIONS OF p

Given: A binomial experiment with two possible outcomes; we call one "success" and the other "failure." Let π equal the probability of success in any one trial of this experiment.

1 Try this experiment n times (where n is large enough to make $n\pi$ and $n(1-\pi)$ both greater than 5).

2 Let p equal the percent of trials that resulted in success. Record p.
3 Now repeat this whole process again to get another value of p.
4 Continue until you have a very large distribution of p's.

Conclusion: Mathematical theory predicts that the distribution of p's will

1 Be *normal* in shape.
2 Have mean equal to π.
3 Have standard deviation equal to

$$\sqrt{\frac{\pi(1-\pi)}{n}}$$

CLASS EXERCISE

5-22 In a firecracker factory quality control is such that 15 percent of the firecrackers made there are duds. Assume you are the inspector. Everyday you collect a random sample of 50 firecrackers from the production line. You light them and count the percent that are duds. You do this every day for 2 years at which time you retire to collect disability income. Your successor collects all your records and draws a graph as we did in Figure 5-12.

a Describe the graph he got. ______

b In this experiment what does π represent? ______

c p? ______

d μ_p? ______

e σ_p? ______

5-23 Suppose you are doing a binomial sampling experiment like the one in Exercise 5-22. Suppose $\pi = 15$ percent.

a What does σ_p represent? ______

b If you collect samples of size 100, what will be the value of σ_p? ______

c If instead you collect samples of size 25, what effect will this have on the value of σ_p? ______ Why? ______

5-24 This experiment illustrates the effect that sample size has on the shape of the sampling distribution of p. Let π equal the true percentage of *all* the words in this book which are four-letter words. Then p will represent the percent of four-letter words in any *sample* of words from this book.

a Have each person in the class pick a page of the book at random.

b Circle the first *10* words on the page (skip numbers and math symbols). You have taken a sample of *size 10*. Write $n = 10$. Then put a check over each four-letter word among those 10.

c Count the number of four-letter words and divide by 10 to find p for this sample. $p =$ ________

d Repeat this process once more for the next 10 words on the page. $p =$ ________

e Repeat this process again for the next 10 words. $p =$ ________

f Summarize the class results in the table shown:

p	Frequency: First Time	Second Time	Third Time	Total Frequency
.00				
.10				
.20				
.30				
.40				
.50				
.60				
.70				
.80				
.90				
1.00				

g Graph the data from part *f* in a histogram.

h Is the shape of your graph somewhat normal? ________

i Now repeat the whole experiment (steps a through g) for samples of size $n = 60$.

BINOMIAL HYPOTHESIS TEST

We can use all the information about the sample distribution of p's to carry out a hypothesis test similar to the ones we did in Chapter 4 using the X^2 statistic. A hypothesis test is like a true-false question. Some hypothesis is given, then we get some sample data, and we use it to decide if the hypothesis is probably true or false. In this chapter we are dealing with hypotheses of the type: "A certain percentage of the population belongs to group 1." In Chapter 4 we dealt with hypotheses of the type: "*B* and *C* are statistically independent."

Now let us formalize the problem we discussed before about Dr. Ketchum's students. The problem was to determine the truth of the professor's claim that 30 percent of students cheated on book reports.

We will assume in our null hypothesis that the professor is correct. Then we will know what kind of results to expect by using the theorem about the sample distribution of p. Then we will take a sample of students and find the percentage who cheated. If it is close to what the professor claims, we will assume his theory is reasonable. But if it is far from what he claims, we will reject his theory. We will use what we know about normal curves to find *critical values of p* which separate values that are close to expected from values that are not close.

SOLUTION TO DR. KETCHUM PROBLEM

1 *Population*

Students who submitted book reports to Professor Ketchum.

2 *Hypotheses*

Let $\pi = P$(a student turns in a book report on an unread book)

$H_0 : \pi = .30$
$H_a : \pi \neq .30$

3 *Sample*

A random sample of 200 students who had written book reports were asked if they had read the book reported on. Of these, 40 admitted to falsifying a book report. 160 said that they had not done so.

$n = 200 \qquad p = 40/200 = .20$

4 *Significance level*

Alpha was chosen to be .01. Since $n\pi = 200(.30) = 60$ and $n(1 - \pi) = 200(.70) = 140$ are both greater than 5, we can use the normal curve. The critical z scores that cut off the middle 99 percent of a normal distribution are $z_c = \pm 2.58$.*

5 *Mean*

$\mu_p = \pi = .30$

* The subscript c on z_c indicates that this is the critical value of z.

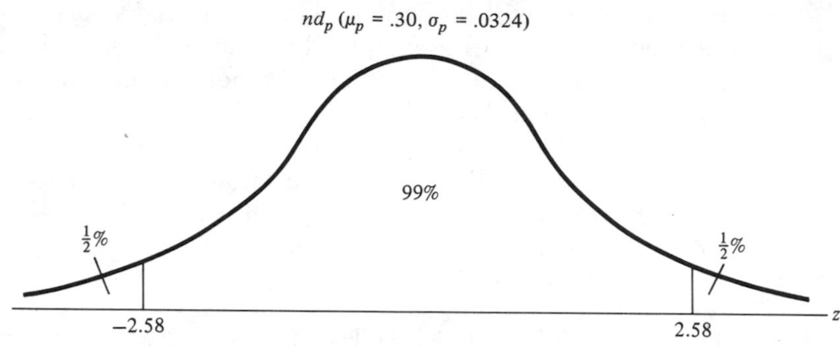

Figure 5-13

6 *Standard deviation*

$$\sigma_p = \sqrt{\frac{\pi(1-\pi)}{n}} = \sqrt{\frac{(.3)(.7)}{200}} = \sqrt{\frac{.21}{200}}$$
$$= \sqrt{.00105} = .0324$$

7 *Compute the critical values*

(See Figure 5-13.)

$$p_c = \mu + z_c\sigma = .30 + (\pm 2.58)(.0324)$$
$$= .30 + (\pm .084) = .216 \text{ and } .384$$

(See Figure 5-14.)

8 *Decision rule*

We have learned from our computations that p will almost always be between 21.6 and 38.4 percent if π is really 30 percent. Therefore, if our sample value of p is not between 21.6 and 38.4 percent, we will decide that π is *not* really 30 percent. When we decide that the null hypothesis is proba-

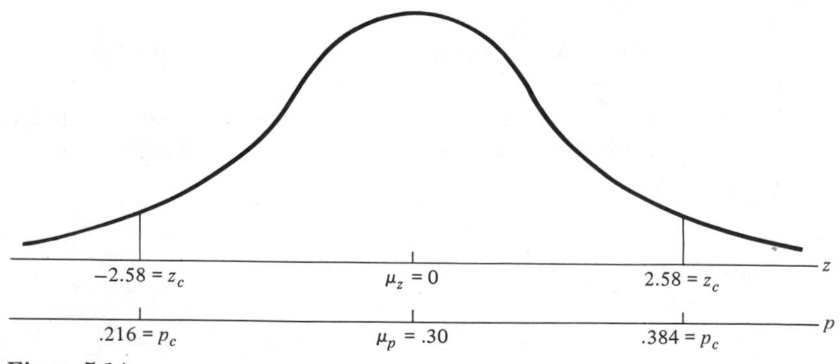

Figure 5-14

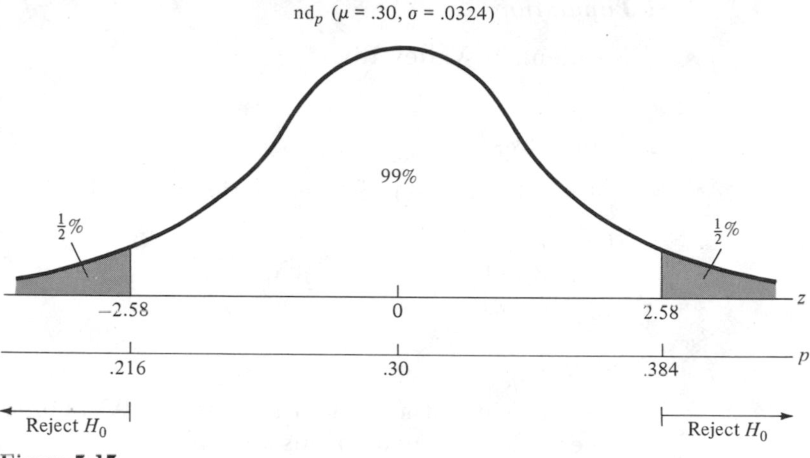

Figure 5-15

bly false, we say formally that we *reject the null hypothesis*. We often indicate this by the type of sketch shown in Figure 5-15.

9 *Outcome*

From step 3 our sample outcome was $p = .20$.

10 *Conclusion*

Since .20 is less than .216, we reject the H_0. π is not equal to .30. In fact we are certain at the .01 significance level that π is less than .30. The professor was in error; 30 percent was too high for the percent of cheaters.

11 *Comments*

Although it is possible that some students may lie about falsifying a book report, the experimenter does not think that this was a significant factor in this hypothesis test. (See Assignment 35.)

Another Example

A newspaper poll states that 38 percent of the population of Metropolis favors the enactment of a stricter drug control law. Suspecting that this figure is not correct in the suburb of Valley River, we decide to do a binomial hypothesis test.

1 Population

Residents of Valley River

2 Hypothesis

Let $\pi = P$(a resident favors a stricter drug law)

$H_0: \pi = .38$
$H_a: \pi \neq .38$

3 Sample

We interviewed a random sample of 203 people in Valley River. 23 declined to answer or were not residents. 80 favored the stricter law. 100 did not. $n = 180$. $p = 80/180 = .44$.

4 Significance level

Let $\alpha = .05$

$$n\pi = 180(.38) = 68.4 > 5$$

$$1 - \pi = .62 \qquad n(1-\pi) = 180(.62) = 111.6 > 5$$

Therefore, we can use the normal distribution to approximate the binomial distribution, and the critical z scores will be ± 1.96.

5 Mean

$\mu_p = .38$

6 Standard deviation

$$\sigma_p = \sqrt{\frac{(.38)(.62)}{180}} = \sqrt{\frac{.2356}{180}} = \sqrt{.00131} = .0362$$

7 Compute critical values

nd_p ($\mu_p = .38$, $\sigma_p = .0362$)

$$p_c = \mu + z_c\sigma = .38 + (\pm 1.96)(.0362)$$
$$= .38 \pm .07 = .31 \text{ and } .45$$

(See Figure 5-16.)

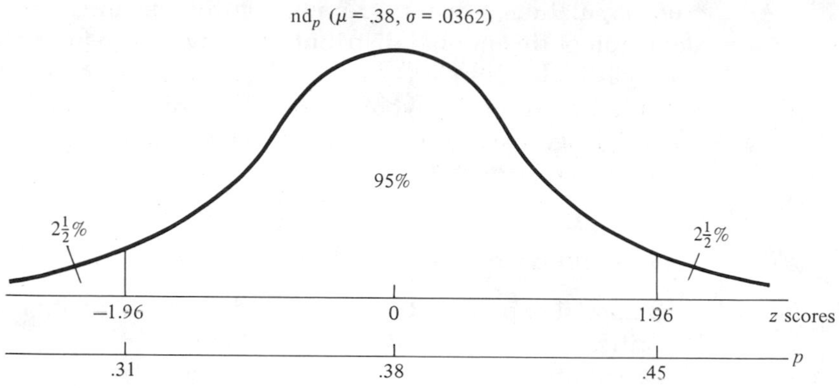

Figure 5-16

8 Decision rule

We will reject the null hypothesis if our sample value of p is less than .31 or greater than .45.

9 Decision

Our experimental result was $p = .44$. Therefore, we fail to reject the null hypothesis. We have not proved at the .05 significance level that the residents of Valley River feel differently from the residents of Metropolis. Even though our outcome (44 percent) was higher than the claim (38 percent), it was not high enough for us to be certain that the 38 percent was incorrect. It is possible that the 38 percent figure is also correct in Valley River. Our computations tell us that it is not that unusual to get a sample value of $p = 44$ percent when the true value of the parameter is $\pi = 38$ percent. The difference may have been due merely to chance factors in the selection of our random sample.

10 Comments

Notice we are not testing the truth of the newspaper article. That article concerned the proportion of residents of Metropolis. We tested the proportion of residents of Valley River.

Chapter Summary

In repeated sampling, a distribution of sample percents tends under certain conditions to be normal with the mean equal to the true population percent. Knowledge of the mean and standard deviation of this normal distribution allows us to judge whether or not a particular sample is likely to have come from the distribution specified by the null hypothesis. The procedure and techniques by which we make this judgment makes up what we call a **binomial hypothesis test.**

Words and Symbols You Should Know

Binomial experiment	Standard deviation	p_c
Critical z score	Statistic	z
Distribution of sample p's	Variability	z_c
Normal curve	z score	μ
Normal distribution	n	μ_p
Parameter	nd_p	π
Range	nd_X ($\mu = 100$, $\sigma = 5$)	σ
Raw score	p	σ_p

ASSIGNMENT

An old statistician at tea
Remarked, Simpson's value of p—
Why it's much too high
To believe that π
Is only point double oh 3

13 A fair die is rolled. Rolling a three is considered success. Rolling anything else is considered failure. Let $\pi = P$(a 3 is rolled) be the definition of π.

a What is the numerical value of π?

b Define $1 - \pi$. $1 - \pi = P(\,?\,)$.

c What is the numerical value of $1 - \pi$?

The die is rolled 100 times and the proportion of threes rolled is recorded as a value of p. This experiment is repeated many times.

d Find μ_p.

e Find σ_p.

f About 95 percent of the values of p should fall between what two numbers?

14 Repeat Assignment 13 using a fair coin. Let success be tossing tails. Toss the coin 80 times and repeat the experiment many times.

15 Identify the type of hypothesis test used to test each of the hypotheses in parts *a* through *e*. Your choices are

(A) chi square (B) binomial (C) neither

a In the U.S. population there is no relationship between the number of children per family and the political affiliation of the parents.

b During the Vietnamese War 30 percent of returning GI's started college.
c The average height of American males is 5 feet $10\frac{1}{2}$ inches.
d A U.S. penny minted in 1973 is a fair coin.
e This pair of dice is fair.

16 *FILE* magazine stated that 60 percent of Americans supported the President's foreign policy. A random sample of students led a researcher to reject the $H_0: \pi = .6$. He then claimed that *FILE* magazine was incorrect. Comment.

Hypothesis tests (see 11-step outline following Exercise 5-24). Be sure to define the population clearly.

17 A coin is tossed 100 times and produces 63 tails. Is this coin fair? Use $\alpha = .01$.

18 A national study at the height of student activism in 1968 claimed that 10 percent of all college students were in serious disagreement with the aims, values, and politics of their parents. If you tested this today on a random sample of 500 students and found 40 students in this category, would this tend to support or deny the 1968 figure? Use $\alpha = .05$.

19 Test the hypothesis that 50 percent of all college students consider themselves to be political moderates or conservatives if a random sample of 300 students has 140 students in this category. Use $\alpha = .05$.

20 A recent Gallup poll showed that 56 percent of Americans would prefer a rural life if they could have it. If a random sample of 1,000 Americans turned up 600 who favored a rural life, would this tend to confirm or deny Gallup's results? Use $\alpha = .05$.

21 A medical scientist predicted on the basis of theory that 50 percent of the people who contracted a certain disease caused by a slow virus would be found to have had measles before the age of 2. A field test on 90 victims of the disease revealed that 40 had measles before age 2. Does this confirm or deny her theory at the .05 significance level?

22 A desperate ad agency predicted that 20 percent of a certain population of insects would have bad breath and so would be a good "target" group for their ads. A field test of 200 randomly captured insects showed 10 with bad breath. At $\alpha = .05$, does this support the prediction?

23 A 4-year study ending in 1973 showed that nationally conservative sexual attitudes predominate in the United States. One question asked of the many people interviewed was this: "Two states have laws which do not forbid sex acts between persons of the same sex, if done in complete privacy between consenting adults. Do you approve or disapprove of such a law?" 60 percent of the people interviewed said they disapproved of such a law. Suppose you tried to compare these results with results obtained on your campus. Suppose you interviewed 100 students and found that 44 disapproved of such a law. What would this prove? Use $\alpha = .05$.

24 In 1967 a national survey of 17-year-old males had claimed that 54 percent of them had stolen property at least once during the time they were 13 to 16 years old. If you took a similar sample today and found that 490 out of 900 had stolen

property, would this lead you to claim that now the percentage was different from what it was in 1967? Let $\alpha = .05$.

25 A student did not believe that if you drew a card from a shuffled deck, you would get diamonds 25 percent of the time. He picked a card 100 times, reshuffling the deck after each draw. He picked 21 diamonds, not 25. Test at the .01 significance level.

26 The manager of Markets Inc. found that 30 percent of his merchandise was delivered defective. The goods were unloaded down a chute. When he had his employees carry the next shipment of 120 iceboxes to an elevator and then down to the warehouse, only 30 pieces were defective. Is the difference statistically significant if $\alpha = .01$?

27 A newspaper stated that 65 percent of all female college students tested at a Midwestern university felt that they would suffer a loss of femininity if they became successful. To see if this figure applied locally, Meg questioned 89 females at her school. Of those questioned 9 were not students; 42 thought that becoming a successful figure would cause a loss of femininity. Do a hypothesis test with $\alpha = .05$.

28 According to national statistics for 1970, 48 percent of all accidental deaths occur in motor vehicle accidents. A student checks in his city and finds that of the last 300 deaths by accident 160 were in motor vehicle accidents. At $\alpha = .05$, does this support the hypothesis that his city is similar to the national scene with regard to accidental deaths?

29 In 1950 the national census showed that 1 percent of the population died that year. A sample of 192,000 people in 1970 showed that 1,800 died that year. At $\alpha = .05$, does this show that the percentage of people dying per year has changed?

30 Two possible cereal boxes (*A* and *B*) for Super Munchies were being compared to see which was more attractive to 8-year-old boys. One at a time, 50 boys were left in a room and told they could try both cereals. It was noted which box they tried first. Box *A* was tried first by 32 boys. The rest tried box *B*. Is this conclusive evidence, at $\alpha = .05$, to suggest that the boxes are not equally attractive?

31 Benny DeKruk was arrested on suspicion of cheating in his gambling casino. Judge Al Ferr, having read "Beginning Statistics," rolled Benny's dice 45 times. He knew that $\pi = P(7 \text{ on a pair of dice})$ is 1/6 if the dice are honest. Noting that $n\pi$ and $n(1 - \pi)$ are both larger than 5, Judge Ferr stated, "We have nd_p ($\mu = 1/6$, $\sigma = 1/18$). Hence, our decision is to find you guilty if p is more than .28 or less than .06."

a Verify the judge's calculations of $n\pi$, $n(1 - \pi)$, μ, σ, and the decision rule which is based on a significance level of $\alpha = .05$.

b If the dice show 15 sevens and the judge finds Benny guilty, Benny should appeal because the judge might have made an error of Type ________ .

c If the dice show 12 sevens, $p = .27$. To be sure of avoiding a Type II error the judge should find Benny:

A Innocent B Not guilty C Guilty

32 A town in California had a special election to try to unseat the mayor because he was very young. At the election about half the eligible voters bothered to vote. The vote was 612 not to unseat and 301 to unseat. Is this convincing evidence ($\alpha = .01$) that the election would have gone the same way if everyone had voted? What assumption must you make about the people who actually did vote? Is it a reasonable assumption? For the mayor's assistant, the vote was 486 not to unseat and 438 to unseat. Answer the same questions for this election.

33 a Get a claim for a binomial percent from any one of these sources: a newspaper, a magazine, a textbook, radio, television, or a teacher.
b Outline a hypothesis test using this number. Be specific as to the population to be tested, the sample to be gathered, the question to be asked, etc.
c Carry out the experiment you have outlined in part b.

34 Make up your own binomial question and use the data in Appendix A to answer it.

35 Read The Juvenile Delinquent Nobody Knows, *Psychology Today*, September 1973 to see how one experimenter tried to handle the problem of people lying during interviews.

Readings from J. Tanur (ed.), "Statistics: A Guide to the Unknown," Holden-Day, San Francisco, 1972.

36 Parking Tickets and Missing Women, Zeisel and Kalven, p. 102. How was a hypothesis test about proportions used to show discrimination in the selection of the jury for the trial of Dr. Benjamin Spock?

37 Information for the Nation from a Sample Survey, Taeuber, p. 285. What is a reinterview program? What does it have to do with the precision of a statement like "13 percent of the U.S. population in 1968 was living in poverty."

38 Measuring Sociopolitical Inequality, Alker, p. 343. This article is a little bit tough to understand. What is a Lorenz curve? How is it used to show that 8 percent of the New York State population elected 20 percent of the State Assembly?

Readings from The New York Times

39 Read about some estimates of percentages from a Gallup Poll concerning tolerance toward sexual behavior, August 12, 1973, p. 21.

Review

40 If the sex of each child born to a family was equally likely to be male or female, then in a family with four children, there would be $2^4 = 16$ ways the boys and girls could be born.
a Using B for boy and G for girl list the 16 possible arrangements in the following chart:

Arrangement	Oldest	Second Born	Third Born	Youngest
1				
2				
3				
4				
5				
6				
7				
8				
9				
10				
11				
12				
13				
14				
15				
16				

b What percentage of the arrangements have 4 boys? 3 boys? 2 boys? 1 boy? No boys?

c In a random sample of 160 families with four children we observed the following data:

8 families with 4 boys
37 families with 3 boys
70 families with 2 boys
39 families with 1 boy
6 families with no boys

Test, with $\alpha = .05$, to see if the sex of each child is equally likely. What is the null hypothesis?

SIX

Two-Sample Binomial Experiments

Vive La Difference
French Adage

Is there a generation gap in politics? More specifically, is it true that younger voters tend to vote for Democrats more than older voters do? This is a big issue in every presidential election. How can we test this hypothesis? One way would be to collect data from *both* a random sample of young voters and a random sample of older voters. Then we could compare the two samples. Such a procedure would be called a **two-sample test.** In this chapter we are going to discuss two-sample binomial tests.

In general we perform a two-sample test when we wish to *compare two populations.* The outline of the test is as follows:

1 State clearly what the two populations are. In the illustration above, the two populations are not defined clearly enough for us to proceed. We might define "young voters" to be all voters 21 years old or younger in the Fifth Congressional District. We might define "older voters" to be all voters in the Fifth Congressional District who are 40 years old or older.
2 State clearly the null and alternative hypotheses. For example, H_0: The percentage of young voters who will vote Democratic *is the same* as the percentage of old voters who will vote Democratic. This one is the null hypothesis because if it is true, we know what to expect in the experiment. Recall that the null hypothesis is always the one which leads to expected results. H_a: The percentage of young voters who will vote Democratic *is not the same* as the percentage of old voters who will vote Democratic.
3 Collect a random sample from *each* population.
4 Decide if the *difference* between the two samples is statistically significant. For example, compare the percentage of Democrats in the two samples to see if there is a significant difference. (We will see how to do this comparison later.)

CLASS EXERCISE

6-1 Consider another question.
Is it true on your campus that the percentage of male students who hold regular part-time jobs (10 hours or more per week) is different from the percentage of female students who hold part-time jobs? Use the males and females in your statistics class as your two random samples. Record your data here:

a Population 1: ______________________

Population 2: ______________________

b Null hypothesis: ______________________

Alternative hypothesis: ______________________

c Size of sample 1: ______

Size of sample 2: ______

d Number in sample 1 with regular part-time job: ______

Number in sample 2 with regular part-time job: ______

e *Percentage* in sample 1 with regular part-time job: ______

Percentage in sample 2 with regular part-time job: ______

6-2 Would you say intuitively that your results in part e support your alternative hypothesis? Remember, the hypothesis is about the entire school, not just your class. ______

COMPARING PERCENTAGES IN TWO RANDOM SAMPLES – COMPUTER SIMULATION

In a two-sample test the computational work usually boils down to deciding whether or not the percentages observed in the two samples are significantly different from each other.

We will examine the basic ideas of this computation by making up an illustration and looking at a computer simulation.

Population 1: Voters, ages 18 to 21, in the Fifth Congressional District.

Population 2: Voters, ages 40 and over, in the Fifth Congressional District.

Null hypothesis: The percentage of voters who intend to vote *Democratic* in the next presidential election *is the same* in both populations.

Alternative hypothesis: The percentage of voters who intend to vote *Democratic* in the next presidential election *is not the same* in both populations.

Now go out and gather the two random samples. Suppose that in a sample of 40 young voters, 30 intended to vote Democratic. In a second sample of 50 older voters, 40 intended to vote Democratic.

If we let $\pi_1 = P$(a young voter will vote Democratic) and $\pi_2 = P$(an older voter will vote Democratic), then $p_1 = 30/40 = .75$ and $p_2 = 40/50 = .80$.

CLASS EXERCISE

6-3 Does this indicate that a larger proportion of older people will vote Democratic?

______

6-4 Is it possible that $\pi_1 = \pi_2$? ______

6-5 Is it likely that $\pi_1 = \pi_2$? ______

Now, *suppose* that the null hypothesis is actually true for the two populations (young and old) in the Fifth Congressional District. Suppose, for example, that 80 percent of the voters in both populations intend to vote Democratic. We are going to look at a random sample from each population.

Since we are supposing that 80 percent of all the young voters intend to vote Democratic, we would expect about 80 percent of our sample of young people to say they intend to vote Democratic. Similarly, we expect about 80 percent of our sample of old voters to say they intend to vote Democratic.

It would be surprising if the two *samples* showed *exactly* the same percentage of Democratic voters even if the two *populations* did have the same percentage. We expect some variation as a part of sampling procedure. Therefore, it will be our job to look at the difference between the two samples and decide if it is the kind of difference we might ordinarily expect from two populations with equal percentages of Democratic voters. This means that we ought to have some way to know what size sample differences are common and what size are uncommon for samples from two populations which each have 80 percent Democratic voters.

In the general case we want to know what differences in sample statistics are usual in the case where the population parameters are actually the same. We let π_1 denote the true parameter in population 1 and π_2 denote the true parameter in population 2. Then p_1 and p_2 are the corresponding sample statistics. We are going to look at the *difference* between p_1 and p_2 (which is $p_1 - p_2$) to help us make a decision about the *difference* between π_1 and π_2 (which is $\pi_1 - \pi_2$).

In order to get to know what size differences to expect, we programmed a computer to contain two large populations: "young voters" and "old voters." In each population we made 80 percent of the "people" Democrats. Then we collected a random sample of 40 "people" from the "young voters" and 50 "people" from the "old voters." Here are the results.

	YOUNG	OLD
Democrats:	$p_1 = 29$ out of $40 = .725$	$p_2 = 39$ out of $50 = .780$

The *d*ifference between these *p*robabilities is denoted by *dp* and equals $p_1 - p_2 = .725 - .780 = -.055$. This shows it is possible for the two samples to differ by 5.5 percent even though the populations have the *same* percentage of Democrats. But is this a usual or an unusual difference? To answer this question we repeated the experiment 4 more times and got the results shown in Table 6-1.

TABLE 6-1

Experiment Number	p_1	p_2	$dp = p_1 - p_2$
1	.725	.780	−.055
2	.700	.820	−.120
3	.800	.740	.060
4	.825	.800	.025
5	.775	.800	−.025

We then ran the program 95 more times and had the computer arrange the 100 values of $p_1 - p_2$ in order of magnitude. This list of numbers is called a **distribution of sample differences.** See Table 6-2.

To see the pattern of outcomes more clearly, we constructed a graph. We first grouped the data as shown in Table 6-3. Then we drew the frequency graph shown in Figure 6-1.

Now we can see what differences to expect in an experiment of this type. We see, for example, that almost all the differences are between −15 and +15 percent. This means that if we did a real experiment of this type comparing a sample of 40 young voters to a sample of 50 old voters, then we should not claim to have proved any difference between these two groups unless our sample difference is more than about 15 percent.

TABLE 6-2

Experiment Number	$dp = p_1 - p_2$	Experiment Number	$dp = p_1 - p_2$	Experiment Number	$dp = p_1 - p_2$
1	−.265	35	−.035	68	+0.35
2	−.180	36	−.030	69	+.040
3	−.155	37	−.030	70	+.040
4	−.150	38	−.025	71	+.045
5	−.150	39	−.025	72	+.045
6	−.145	40	−.025	73	+.050
7	−.140	41	−.020	74	+.055
8	−.135	42	−.020	75	+.060
9	−.130	43	−.020	76	+.060
10	−.120	44	−.020	77	+.070
11	−.110	45	−.020	78	+.070
12	−.110	46	−.015	79	+.075
13	−.105	47	−.015	80	+.075
14	−.105	48	−.010	81	+.075
15	−.090	49	−.010	82	+.080
16	−.090	50	−.010	83	+.080
17	−.080	51	−.010	84	+.085
18	−.080	52	−.010	85	+.085
19	−.075	53	−.005	86	+.085
20	−.070	54	−.005	87	+.095
21	−.065	55	+.005	88	+.105
22	−.065	56	+.005	89	+.105
23	−.060	57	+.010	90	+.115
24	−.060	58	+.015	91	+.115
25	−.055	59	+.015	92	+.115
26	−.050	60	+.020	93	+.115
27	−.050	61	+.020	94	+.130
28	−.050	62	+.020	95	+.135
29	−.045	63	+.020	96	+.135
30	−.045	64	+.025	97	+.140
31	−.045	65	+.025	98	+.165
32	−.040	66	+.035	99	+.180
33	−.040	67	+.035	100	+.285
34	−.035				

TABLE 6-3

Interval	Frequency
−.27 to −.21	1
−.21 to −.15	2
−.15 to −.09	11
−.09 to −.03	21
−.03 to .03	30
.03 to .09	21
.09 to .15	11
.15 to .21	2
.21 to .27	0
.27 to .33	1

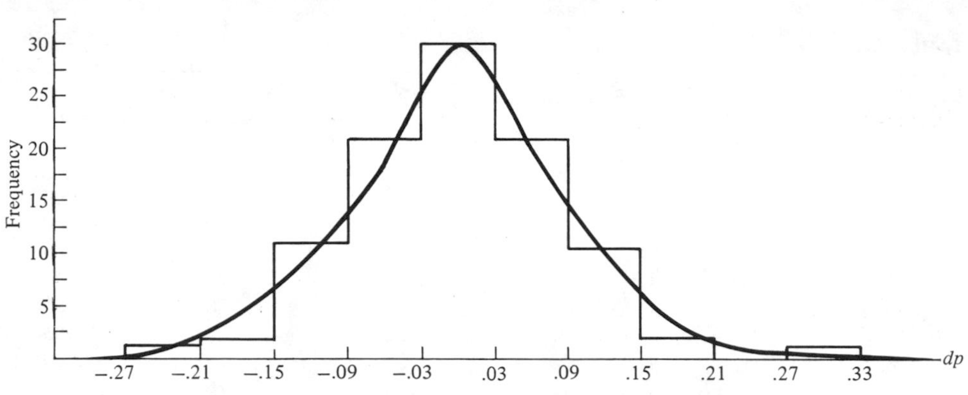

Figure 6-1

COMPARING PERCENTAGES (THEORY)

You can notice that as we smooth out the graph in Figure 6-1, we get a curve which resembles a normal curve. This is not an accident. It can be shown by advanced mathematics that a large distribution of sample differences (if we keep taking more and more pairs of samples) is normal provided that the four numbers, $n_1\pi_1$, $n_1(1-\pi_1)$, $n_2\pi_2$, and $n_2(1-\pi_2)$ are each greater than 5. The mean of this distribution equals $\pi_1 - \pi_2$. Since our null hypothesis assumes that $\pi_1 = \pi_2$, this mean will be zero. The standard deviation of these differences is given by

$$\sigma_{dp} = \sqrt{\frac{\pi(1-\pi)}{n_1} + \frac{\pi(1-\pi)}{n_2}}$$

where $\pi_1 = \pi_2 = \pi$.

In most experiments, of course, π_1 and π_2 are unknown parameters and, furthermore, in most experiments we have only *one* pair of samples. In this case, we will use p as an estimate of π. We compute p as follows.

Let f_1 equal the number of successes observed in population 1 in n_1 observations. Then

$$p_1 = f_1/n_1$$

Let f_2 equal the number of successes observed in population 2 in n_2 observations. Then

$$p_2 = f_2/n_2$$

Since we are assuming that $\pi_1 = \pi_2$, we can combine our two samples and consider them as one larger sample from the same population. Thus, we can estimate the value of π by

$$p = \frac{f_1 + f_2}{n_1 + n_2}$$

That gives us

$$s_{dp} = \sqrt{\frac{pq}{n_1} + \frac{pq}{n_2}}$$

as an estimate of σ_{dp}. Finally, we check that n_1p_1, n_1q_1, n_2p_2, n_2q_2 are all greater than 5.

Illustration

This illustration is based on medical reports published in 1973 which claimed that women who use birth control pills and women who do not are not equally likely to have heart disease. What is the statistical procedure for coming to such a conclusion?

PROCEDURE

1 Identify populations

Population 1: Women who use birth control pills.
Population 2: Women who do not use birth control pills.

2 State hypotheses

Let $\pi_1 = P$(a woman has heart disease in population 1).
Let $\pi_2 = P$(a woman has heart disease in population 2).
Null hypothesis: There is no difference between the percentages of women in these two populations who have heart disease. $\pi_1 = \pi_2$ or $\pi_1 - \pi_2 = 0$.
Alternative hypothesis: There *is a difference* between the percentages of women in these two populations who have heart disease. $\pi_1 - \pi_2 \neq 0$.

3 Choose significance level

$\alpha = .05$

4 Observe data

Random sample of 400 women from population 1 reveals 30 cases of heart disease.

$p_1 = 30/400 = .075 \qquad q_1 = 370/400 = .925$

Random sample of 300 women from population 2 reveals 10 cases of heart disease.

$p_2 = 10/300 = .033 \qquad q_2 = 290/300 = .967$

Pooling these sample results, we get

$$p = \frac{30 + 10}{400 + 300} = \frac{40}{700} = .057 \qquad q = .943$$

5 *Introduce normal curve*

A Check n_1p_1, n_1q_1, n_2p_2, and n_2q_2.

$$n_1p_1 = (400)\ \tfrac{30}{400} = 30$$
$$n_1q_1 = (400)\ \tfrac{370}{400} = 370$$
$$n_2p_2 = (300)\ \tfrac{10}{300} = 10$$
$$n_2q_2 = (300)\ \tfrac{290}{300} = 290$$

These four numbers are exactly the same as the experimenter's results: 30 women using the pill had heart disease, 370 did not; 10 women not using the pill had heart disease, 290 did not. This is always the case in a two-sample binomial hypothesis test.

B Since 30, 370, 10, and 290 are all greater than 5, we can assume that the distribution of sample differences $p_1 - p_2$ will be normal, with mean equal to $\pi_1 - \pi_2$, $z_c = \pm 1.96$, and

$$s_{dp} = \sqrt{\frac{pq}{n_1} + \frac{pq}{n_2}}$$

C Since our null hypothesis claims $\pi_1 = \pi_2$, we then expect $\pi_1 - \pi_2$ to equal 0. We compute

$$s_{dp} = \sqrt{\frac{(.057)(.943)}{400} + \frac{(.057)(.943)}{300}}$$
$$= \sqrt{.0001344 + .0001792}$$
$$= \sqrt{.000314} = .0177.$$

We construct a normal curve with mean equal to 0 and standard deviation equal to .0177. (See Figure 6-2.)

6 *Formulate decision rule*

Critical scores $= \mu + z_c\sigma = 0 + (\pm 1.96)(.0177) = \pm .0347$.
Decision rule: If our observed dp is more than .0347 or less than $-.0347$, we will reject the null hypothesis and conclude that the two populations have *different* percentages of women with heart disease. This is because values of dp

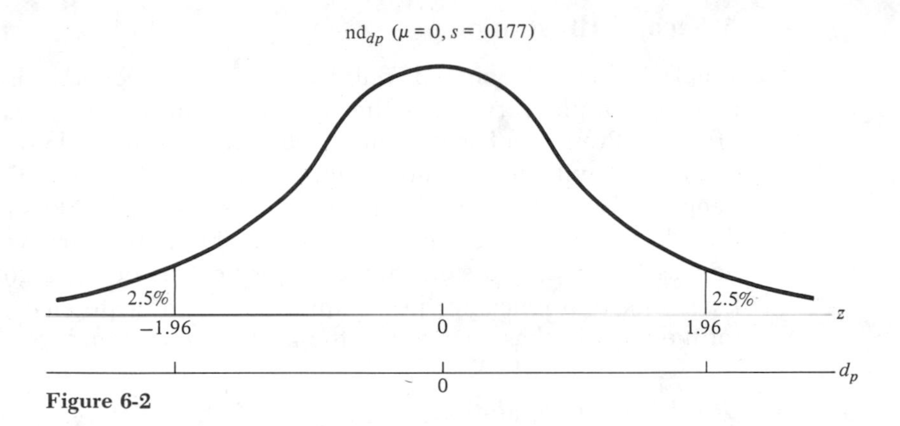

Figure 6-2

between −.0347 and +.0347 frequently are observed even for two populations where $\pi_1 = \pi_2$. That is what the z scores ±1.96 tell us. Conversely, values of $p_1 - p_2$ outside of the range −.0347 to +.0347 hardly ever are observed when two populations have $\pi_1 = \pi_2$. Therefore, if we observe dp outside the range −.0347 to +.0347, we will conclude that probably $\pi_1 \neq \pi_2$.

7 *Conclusion*

The observed values of p_1 and p_2 were .075 and .033. Therefore, the observed value of $dp = .075 - .033 = .042$ is greater than the value in the decision rule. Hence, we conclude that the null hypothesis is probably false. There is evidence that women who use the pill are more likely to have heart disease.

8 *Comments*

Assuming that the medical data was accurate and meaningful, we have statistical evidence (at the .05 significance level) that a greater percentage of women using birth control pills have heart disease than women not using such pills. However, it must still be decided by any given woman and her doctor whether this difference is clinically significant. This of course depends on the risks involved and on other personal matters. The point is that the statistics only show that the *difference exists.* It must be decided on other grounds whether or not this difference is actually important to any particular individual.

A Second Illustration

Crank Galoot, the noted public opinion surveyor, conducted random sample surveys of the population of Weuns *before and after* the Premier of that country was accused of hanky panky in connection with the country's agricultural production. Of 3,000 people surveyed *before* the Premier was accused, 600 said they thought he was doing a good job. Of 2,900 people surveyed *after* he was accused, 506 said they thought he was doing a good job. Does this difference in the sample results indicate that support for the Premier has changed in the *entire* population?

1 *Identify populations*

Population 1: The citizens of Weuns *before* the accusations.
Population 2: The citizens of Weuns *after* the accusations.

2 *State hypotheses*

Let $\pi_1 = P$(a Weun supports the Premier before the accusation).
Let $\pi_2 = P$(a Weun supports the Premier after the accusation).
Null hypothesis: The percentage of citizens supporting the Premier *before* the accusations, π_1, *is the same* as the percentage supporting him after the accusations, π_2. Therefore, $H_0 : \pi_1 - \pi_2 = 0$.
Alternative hypothesis: The percentage of support *is not the same* before and after. $\pi_1 - \pi_2 \neq 0$.

3 *Choose significance level*

$\alpha = .01$.

4 *Observe data*

$$p_1 = \frac{600}{3{,}000} = .200 \qquad q_1 = .800$$

$$p_2 = \frac{506}{2{,}900} = .174 \qquad q_2 = .826$$

$$p = \frac{600 + 506}{3{,}000 + 2{,}900} = .187 \qquad q = .813$$

5 *Introduce normal curve*

$$n_1 p_1 = 600 > 5 \qquad n_1 q_1 = 2{,}400 > 5$$
$$n_2 p_2 = 506 > 5 \qquad n_2 q_2 = 2{,}394 > 5$$

All are greater than 5. Therefore, assume that distribution of $p_1 - p_2$ is normal with mean equal to $\pi_1 - \pi_2$ and

$$s_{dp} = \sqrt{\frac{pq}{n_1} + \frac{pq}{n_2}}$$

Since the null hypothesis claims $\pi_1 - \pi_2 = 0$, the mean $= 0$.

$$s_{dp} = \sqrt{\frac{(.187)(.813)}{3{,}000} + \frac{(.187)(.813)}{2{,}900}} = \sqrt{.00005068 + .00005242}$$
$$= \sqrt{.0001031} = .0102.$$

Therefore, we have

nd_{dp} ($\mu = 0$, $s = .0102$)

6 *Formulate decision rule*

Critical scores: since $\alpha = .01$, $z_c = \pm 2.58$

$dp_c = \mu + z_c s = 0 + (\pm 2.58)(.0102) = \pm .0263$

(See Figure 6-3.)
Decision rule: If our observed dp is outside the range $-.0263$ to $+.0263$, we will reject the null hypothesis. We will conclude that the populations are *not* alike. We will conclude that support for the Premier changed.

7 *Conclusion*

$dp = .200 - .174 = .026$, and .026 is *not* outside the range $-.0263$ and $+.0263$. Therefore, we cannot say that we have

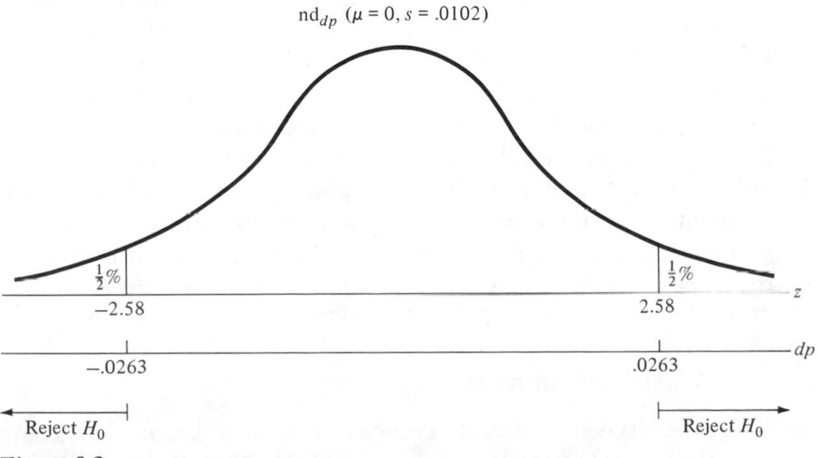

Figure 6-3

proof *at the .01 significance level* that he has lost support *in the entire population.*

8 Comments

You see how close the observed *dp* came to the critical score for the decision rule. This may lead you to want to say that you have "almost" proved that the null hypothesis is false. You must remember, however, that it was the choice of α that determined the critical z scores. If you do this same experiment with $\alpha = .05$ instead of $\alpha = .01$, you *would* be able to claim to have proved the null hypothesis false *at the .05 significance level.* Remember that in sampling statistics you never have absolute proof. If you use $\alpha = .05$ instead of $\alpha = .01$, then there is a greater chance for you to come to a wrong conclusion. There is a greater chance that the variation introduced by random sampling will lead you to reject a null hypothesis which is actually true.

Whether an experimenter chooses to use $\alpha = .01$, $\alpha = .05$, or some other number depends on several things, but one of the most important factors in this choice is: What is the risk or penalty for rejecting a null hypothesis which is actually correct? Are lives at stake? Is a great deal of money at stake? Are people's jobs at stake? The bigger the risk the smaller you want α.

CLASS EXERCISE

6-6 (*Note to the Instructor: You will need a large supply (about 1,000) of identically shaped poker chips and two large bags for this experiment. Some but not all of the chips should be red.*) Mix some poker chips thoroughly in one paper bag. Have someone secretly remove about half the chips and put them into the other bag, either randomly or by careful selection, any way he wants. By drawing a random sample of about 40 chips from each bag, test the hypothesis that both bags contain the same percentage of red chips.

6-7 Select two different textbooks from students in this class. From each book select about 50 lines at random and record the number of words in each line which have more than 5 letters. Test the hypothesis that the two books have the same percentage of words with more than five letters.

Chapter Summary

We have seen how to compare two populations by taking a sample from each one and testing for a significant difference between the two sample statistics. In this chapter the statistics we compared were sample percents. In a later chapter we will compare sample averages. The basic theoretical concept is that under certain conditions distributions of sample differences of percents are normal with mean equal to the true difference between the population percents.

Words and Symbols You Should Know

Distribution of sample differences
Sample difference
Two-sample hypothesis test

dp	p_2	μ_{dp}
n_1	p	π_1
n_2	q_1	π_2
nd_{dp}	q_2	σ_{dp}
p_1	q	

ASSIGNMENT

A chef made a cake and a pie too
To compare his π_1 and his π_2
He preferred π_1
In his test the pie won
Now whoever shall he give the pie to?

1 Choose which type of statistical analysis would be used to answer each of the Questions a through e below. Make your choices from:

A Chi square B Binomial one-sample C Binomial two-sample
D Other

a Do 30 percent of Martians glarph?
b Do the same percentage of Martians and Venusians glarph?
c Is the average height of Martians 3 feet 6 inches?
d Are Venusians taller than Earthlings?
e Is there any relation between height and planet of origin?

2 Why would you not use the methods of this chapter to test whether the percentage of boys who do differs from the percentage of girls who do, if in a random sample of 50 girls, 30 do while in a random sample of 50 boys, 48 do?

3 Find a random sample of about 50 males. Ask each one (so that no one else hears the response) to name a number from 1 to 10. Repeat this procedure with about 50 females. Test the following hypothesis: In an experiment of this type the percentage of females naming a number less than 5 is the same as the percentage of males naming a number less than 5. Use $\alpha = .05$.

4 a For the data in Table 6-2, the actual mean of the 100 computer differences was −.0015 and the standard deviation came out to be .089. Compute the theoretical mean and the theoretical standard deviation by referring to the original problem preceding the table.
  b Do a hypothesis test using the sample data given just before Exercise 6-3. Test at the .01 significance level.
5 Spin a coin on a table 25 times. Record the number of heads. Then toss the same coin in the air, catch it, and slap it down on your wrist. Record the number of heads. Repeat this also for a total of 25 tosses. Is there evidence that the percentage of heads is different for these two techniques?
6 How would you design an experiment to test this hypothesis: Children with reading disabilities are more likely than normal readers to be subject to cruel treatment in the classroom?
7 Dr. Romax has invented a marvelous new potion for making people supercreative. The potion is given to a random sample of 60 left-handed people; 40 become supercreative. The potion is given to a random sample of 80 right-handed people; 60 became supercreative. Does this indicate that the potion works differently on right- and left-handed people? ($\alpha = .05$)
8 A criminologist was studying the responsiveness of persons who witness a crime. At the beach an experimenter left a portable radio on a blanket near a person selected at random and then left. A short time later a second experimenter walked off with the radio. This experiment was repeated 40 times. The witness intervened to stop the "crime" 8 times. On another day the experiment was repeated, but this time the witness was asked to keep an eye on the radio while the experimenter went away. On this day the "crime" was stopped 34 out of 40 times. Show how the statistics indicate that there is a significant difference between the two days. What are the two populations being compared? Let $\alpha = .01$.
9 Dr. J. Van Ghent, the noted Belgian scientist, raised two samples of nematode worms. All the nematodes were fed heme, an iron-containing pigment of hemoglobin; 500 received heme in a solid state; 400 nematodes received heme as a liquid. In the former group, 400 remained alive while in the latter sample 300 survived. Does this support the theory that the heme acts differently in the liquid and solid states? Use the .01 significance level.
10 A random survey of 400 women who regularly smoke cigarettes showed 100 with "abnormal" lung tissue. A random sample of 600 women who do not smoke showed 50 with "abnormal" lung tissue. With $\alpha = .05$, is there a significant difference between the two samples? Explain why even a significant difference would not *by itself* prove that smoking *causes* an increase in abnormal lung tissue.
11 Is there any difference between the attitude of parents and the attitude of nonparents toward more lenient laws regarding the possession of marijuana? In the town of Centerville, a random sample of 80 parents was gathered and 30 favored a relaxation of the law. A sample of 50 nonparents showed that 40 favored more lenient laws. Test at the .05 significance level.
12 To see if white men react differently to black persons than to white persons

asking about civil rights, a population of white males was sampled twice. In the first sample white interviewers asked, "Do you think that the civil rights movement has gone too far?" Out of 498 persons surveyed, 98 refused to cooperate or were vague, 120 said yes, 280 said no. In the second sample, black interviewers asked the same question. Of 300 people responding this time, 60 said yes and 240 said no.

a What are the populations being sampled?

b Using $\alpha = .01$, do a hypothesis test on their responses.

13 Two perfumes were being tested to see if they were equally effective in attracting men. A pair of identical twins (female) were "done up" identically, except for their perfume. A man was selected at random and introduced to one of the females. By a sophisticated electronic device it was determined if he was attracted to her. This experiment was repeated until 100 men had been introduced to female A and 100 had been introduced to female B. It was found that 15 were attracted to A, and 20 were attracted to B. Is this sufficient at $\alpha = .05$ to establish that one perfume is more effective than the other?

14 On Sunday June 2, 1974 an ad appeared on page 6 section E of *The New York Times*. A highly respected opinion research firm asked 1,653 people the following question: "The United States Supreme Court has ruled that a woman may go to a doctor to end pregnancy at any time during the first 3 months of pregnancy. Do you favor or oppose this ruling?" The following week they sampled another 1,653 people. Only this time the words "to end pregnancy" were changed to "for an abortion." The results were as follows:

Interview	Favor	Oppose	Refused to Answer	Undecided
First	45.9%	39.5%	10.0%	4.6%
Second	41.0%	48.7%	5.9%	4.4%

a Test to see if there is any significant difference between the percentage of people who favored the ruling in the first poll and the percentage who favored the ruling in the second poll. Use $\alpha = .01$. (*Hint:* To find p, first find how *many people* favored the ruling in each sample.)

b A week later a third sample took place using exactly the same wording as the second. This time 1,652 people were interviewed. The results were:

Interview	Favor	Oppose	Refused	Undecided
Third	42.9%	53.7%	2.3%	1.1%

Convert the percentages in the second and third interviews back to the *number* of people responding. Then perform a chi-square test on the results to see if there is a relation between which interview people participated in and what their opinions were. In terms of the problem, what is the hypothesis of independence? What will it mean if the sample value of X^2 is significantly large?

15 A group of ecologists was studying the manner in which moths reproduce. In particular, they were studying how the male moths of one species (call it species A) manage to find the female of the correct species, since the females of other species are similar. They found that in several species the females use a mixture of the same two chemicals as the sexual attractant, but the chemicals are mixed in different ratios in different species. When they loaded a trap with one ratio of the chemicals, they caught 190 type A moths and 20 other moths. When they loaded a trap with another ratio of the chemicals, they caught 30 type A moths and 160 other moths. Interpret this data by a two-sample binomial test with $\alpha = .01$. Test the hypothesis that males of type A respond similarly to different mixes of the attractant chemicals. What are the two populations for the test? What will π_1 represent?

16 a Outline a two-sample binomial hypothesis test. State clearly the binomial question, the two populations, the sampling procedure, etc.

b Carry out the experiment that you designed in part a.

17 Make up your own question and use the data in Appendix A to answer it.

Readings from J. Tanur (ed.), "Statistics: A Guide to the Unknown," Holden-Day, San Francisco, 1972.

18 The Biggest Public Health Experiment Ever: The 1954 Field Trial of the Salk Poliomyelitis Vaccine, Meier, p. 2. How convincing was the difference in percent of polio cases between the control group and the vaccinated group? What is a double-blind experiment?

19 Deciding Authorship, Mosteller and Wallace, p. 164. How was a comparison of percents used to decide which of *The Federalist* papers were authored by Alexander Hamilton and which by James Madison?

20 Safety of Anesthetics, Moses and Mosteller, p. 14. Comparing percentages of patients who die while under different anesthetics is not easy. How are some of the complications overcome?

Readings from The New York Times

21 Read in the October 21, 1973 issue, on p. 23, about some differences in jury verdicts due to pretrial news coverage.

Review

22 A guidance counselor at Random High School wants to see if there is any connection between students' smoking and drinking and their class levels. He interviewed 1,000 students and accumulated the following data:

Class	Drink and Smoke	Drink Only	Smoke Only	Neither
Freshman	70	80	40	60
Sophomore	80	20	70	80
Junior	110	20	60	60
Senior	100	40	30	80

Do these data indicate that the drinking and smoking habits of students are related to the class they are in? What is the null hypothesis? Use $\alpha = .01$.

23 Do 10 percent of parents watch the reruns of the Mickey Mouse Club on TV if a random sample of 400 parents showed that 20 watched? Use the .01 significance level.

SEVEN

Averages and the Central-Limit Theorem

The average age of dwellings has steadily declined.

*E. F. Carter, Stanford Research Institute**

* Quoted in Alvin Toffler, "Future Shock," Bantam Books, New York, 1971, p. 55.

Many important statistical problems are about *averages*. For example, an insurance company might want to know the average life span for the American male; a guidance counselor might want to know the average score that high school seniors get on the college board examination; a union official might want to know the average weekly income for construction workers; a government economist might want to know the average amount of money a U.S. family spends on food every month.

CLASS EXERCISE

7-1 Give an example, different from those above, of an important use of averages.

Why is an average useful? Primarily because it is *one* number that stands for or represents a whole collection of numbers. You can remember this one value instead of a long list of values. For example, if you hear that the average temperature in New York City in August is 90°, then you accept that as representative. If you go to New York in August, you expect the temperature to be about 90°. Sometimes it is not fair just to use an average to represent a longer list, but often it *is* fair and reasonable. Here are two examples:

1 Leonard took a math course where he had an exam every month for 5 months. Here are his scores:

Sept.: 20 Oct.: 40 Nov.: 60 Dec.: 80 Jan.: 100

$$\text{Average} = \frac{20 + 40 + 60 + 80 + 100}{5} = \frac{300}{5} = 60$$

Conclusion: Leonard's average test score is 60. He deserves a D in the course.

2 Ellen is a buyer for a hardware store. She wanted to know the average number of hours that a Xeno battery lasts when it is used in a new kind of flashlight. She set up five flashlights and let them burn until the batteries were worn out. Here are her results:

Light 1: 190 minutes Light 2: 193 minutes
Light 3: 180 minutes Light 4: 187 minutes
Light 5: 191 minutes

$$\text{Average} = \frac{190 + 193 + 180 + 187 + 191}{5} = \frac{941}{5} = 188.2 \text{ minutes}$$

Conclusion: Xeno batteries last an average of 188.2 minutes in these flashlights.

CLASS EXERCISE

7-2 In which example above is it fairer to use the average as representative of the list of results? ________ Why? ________________________

7-3 Make up an example where it would be fair to use an average. Make up some appropriate numbers for the problem and actually compute the average.

____

7-4 Make up an example where it would *not* be fair to use an average. Make up some appropriate numbers, compute the average, and show *why* the average is not a fair representation of the numbers. ________________________

DIFFERENT KINDS OF AVERAGES

The ordinary familiar average where you add up all the values in the list and then divide by the *number* of values in the list is called the **arithmetic mean.** To many people the word "average" or the word "mean" automatically denotes this arithmetic mean. In this book we will use the mean more than any other kind of average.

CLASS EXERCISE

7-5 Find the mean of these sets of numbers.

a 1, 1, 1, 1 ________________________

b 1, 0, 0, 2, 3 ________________________

c 42, 47, 49, 49, 49 ________________________

d 22, 27, 29, 29, 29 ________________________

e 0, 3, 2, 2, 3, 4, 7 ________________________

f 1,000,002; 1,000,008; 1,000,010 ________________________

In an experiment, a random sample of 500 people is chosen from the population of the members of ROTC. The mean height of the people in the sample was computed to be 5 feet 10 inches.

7-6 Can we infer that the mean of the entire population of ROTC members is 5 feet 10 inches? ____________

Why? ____________

7-7 Can we infer that the mean of the entire population is definitely close to 5 feet 10 inches? ____________

Why? ____________

7-8 Can we infer that the mean of the entire population is probably close to 5 feet 10 inches? ____________

Why? ____________

What is an average? It is a number that measures in some way the general trend of a distribution of numbers. Besides the arithmetic mean there are the median, the mode, the midrange, and many more.

The **median.** Just as a median on a highway is in the middle of the highway, the median of a list of numbers is in the middle of the list. (Of course, the values have to be *in* numerical order before you can tell which value is the middle one.)

Example

Find the median of 3, 1, 2, 1, 4, 7, 8.

a Put in order: 1, 1, 2, 3, 4, 7, 8

b Find middle one: 3
Since there are seven values, the fourth one is the middle one.
The median is 3.

CLASS EXERCISE

7-9 Find the median for these sets of numbers.

a 1, 1, 1, 1, 1 ____________

b 1, 0, 0, 2, 3 ____________

c 42, 47, 49, 49, 49 ____________

d 22, 27, 29, 29, 29 ____________

e 0, 3, 2, 2, 3, 4, 7 ____________

7-10 Make up a set of five different values where the mean and median are equal to each other. ______

7-11 Make up a set of five values where the mean and median are *not* equal to each other. ______

Example

Find the median of 15, 21, 17, 30, 29, 25.

a Put in order: 15, 17, 21, 25, 29, 30

b Since the list contains an even number of values (six), there is no single value in the middle. Therefore, find the *two middle* values: 21, 25.

c Find the mean of these two.

$$\frac{21 + 25}{2} = \frac{46}{2} = 23$$

The median is 23.

In choosing which kind of average to use, the statistician should choose the one which seems to be more fair, that is, more representative of the original list of values.

CLASS EXERCISE

7-12 Very often a distribution of salaries is very unsymmetric. Such a distribution may consist of a few very high salaries but many which are much lower. Which type of average do you think is most representative when there are a few extreme values? ______

7-13 The salaries in a small business were as follows:

Boss: $40,000
Employee *A*: $6,000
Employee *B*: $6,000
Employee *C*: $8,000
Employees *D* and *E*: $8,500

What is the mean salary? ______

What is the median salary? ______

Which is more representative? ______

The **mode** is the number that appears most frequently. The mode of 3, 1, 3, 4, 8, 2, 3, 4 is 3. The modes of 1, 2, 4, 2, 7, 9, 2, 3, 4, 4 are 2 and 4. 1, 3, 9 has no mode (or equivalently, every number is a mode).

The **midrange** is the arithmetic mean of the smallest and largest numbers in a distribution. In the three examples above, the midranges are

$$\frac{1+8}{2} = 4.5 \qquad \frac{1+9}{2} = 5 \qquad \frac{1+9}{2} = 5$$

CLASS EXERCISE

7-14 Suppose that you were looking into a jeweler's window. Six clocks were on display. They indicated that the time was 6:04, 6:04, 6:10, 6:06, 8:02, and 6:04. Which average do you think would get you closest to the correct time: the mean = 6:25, the median = 6:05, the mode = 6:04, or the midrange = 7:03?

____

The Average of a Sample

Dr. Jerry Attrick claims that the average weight of American males over 60 years old is 150 pounds. You gather a random sample of 36 such people. You compute the mean of these 36 weights, and the result is 155 pounds.

CLASS EXERCISE

7-15 Does this sample mean of 155 pounds tend to make you believe the claim that the average weight of the population is 150 pounds, or does it tend to make you reject the claim as false? ______________________________

7-16 Does the sample size seem large enough to answer the above question? ______

____

INTRODUCTION: HYPOTHESIS TESTS ABOUT AVERAGES

In the previous chapters the problems have been about *percents*. We asked questions like, "Is it true that 30 percent of the population has fleas?" or "Is it true that the same percentage of Americans and Russians distrust the Chinese?" There are many other problems, however, that do not deal with percents. One type deals with *averages*. We ask questions like, "Is it true that automobile batteries last an average of 25 months?" or "Is it true that the average number of inches of snow per year in Seattle is the same as the average number of inches of snow per year in Philadelphia?" In this chapter we are going to discuss some hypothesis tests about averages.

COMPUTER SIMULATION

Once again, you see that it would be helpful to know which sample results are common for samples from a specified population. With regard to the particular claim by Dr. Attrick, we would like to know which sample averages (*sample means*) are common for random samples of 36 weights from a population with mean equal to 150 pounds. We will use the computer to generate a **distribution of sample means.** We wrote a program to select 36 random weights from a population whose mean was exactly 150 pounds. The computer did this 3 times and the results were as follows:

Sample 1

127.1	150.0	179.6	167.8	158.4	151.4
140.8	152.4	156.9	144.4		
150.0	144.4	135.3	170.9	139.9	146.4
145.0	154.3	162.8	161.0		
139.5	156.2	132.2	151.2	151.4	159.9
139.8	158.5	172.4	127.3		
157.8	146.7	135.7	180.7	158.9	156.9

Mean of sample 1 is 151.8 pounds. M = 151.8

Sample 2

133.4	143.6	156.0	152.5	158.0	157.5
161.7	155.5	142.4	145.5		
167.8	129.2	158.5	149.8	147.2	154.8
159.4	153.4	152.7	156.2		
137.4	138.0	136.3	135.3	144.6	166.4
150.5	134.0	134.2	161.7		
148.1	155.5	138.2	125.6	141.6	142.7

Mean of sample 2 is 147.9 pounds. M = 147.9

Sample 3

149.1	160.6	143.3	156.9	159.4	170.1
145.8	161.7	161.8	155.7		

161.5	158.9	149.0	120.3	162.9	111.5
133.3	146.0	122.8	141.4		
147.0	155.9	153.3	151.5	142.7	172.1
147.5	164.3	152.8	157.4		
156.9	142.7	130.3	154.0	154.9	128.4

Mean of sample 3 is 149.5 pounds. M = 149.5

These three sample means are all very close to the population mean which was 150 pounds.
The computer repeated this procedure of picking 36 random weights for a total of 100 times. We then had the computer arrange the results in numerical order. The resulting distribution of sample means is in Table 7-1. If we graphed these 100 numbers (see Table 7-2), we would get the graph shown in Figure 7-1. A smooth curve sketched through this graph has a bell shape to it (see Figure 7-2).

TABLE 7-1

Numerical Order	Mean	Numerical Order	Mean	Numerical Order	Mean
1	144.8	35	148.8	68	150.8
2	144.9	36	148.8	69	150.9
3	145.5	37	148.9	70	150.9
4	145.6	38	149.0	71	150.9
5	146.2	39	149.0	72	151.0
6	146.3	40	149.1	73	151.1
7	146.0	41	149.2	74	151.1
8	146.5	42	149.3	75	151.1
9	146.9	43	149.3	76	151.1
10	147.0	44	149.4	77	151.1
11	147.0	45	149.4	78	151.3
12	147.2	46	149.5	79	151.3
13	147.3	47	149.5	80	151.4
14	147.4	48	149.5	81	151.4
15	147.5	49	149.5	82	151.4
16	147.5	50	149.0	83	151.4
17	147.5	51	149.9	84	151.4
18	147.7	52	149.9	85	151.6
19	147.8	53	149.9	86	151.7
20	147.9	54	149.9	87	151.8
21	148.0	55	150.1	88	151.9
22	148.1	56	150.2	89	151.9
23	148.1	57	150.2	90	152.0
24	148.2	58	150.4	91	152.1
25	148.2	59	150.5	92	152.1
26	148.3	60	150.5	93	152.4
27	148.4	61	150.5	94	152.5
28	148.4	62	150.5	95	152.8
29	148.4	63	150.5	96	153.0
30	148.5	64	150.7	97	153.1
31	148.5	65	150.7	98	153.1
32	148.5	66	150.8	99	154.0
33	148.7	67	150.8	100	154.6
34	148.8				

TABLE 7-2

Interval	143–145	145–147	147–149	149–151	151–153	153–155
Frequency	2	7	28	34	24	5

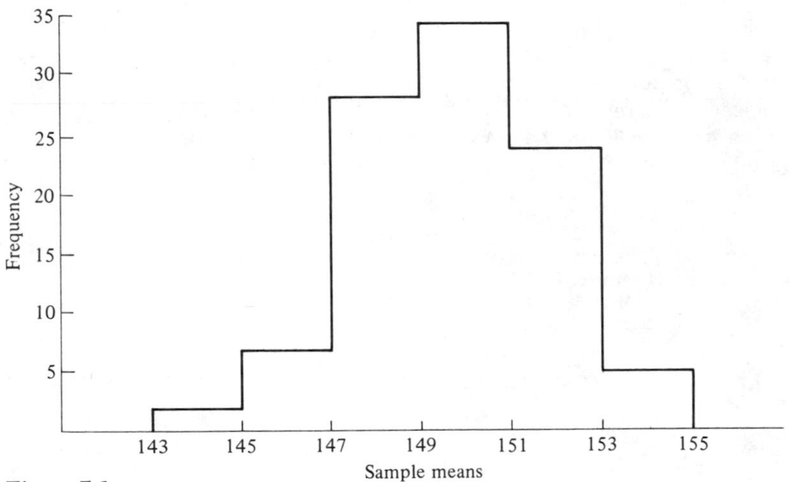

Figure 7-1

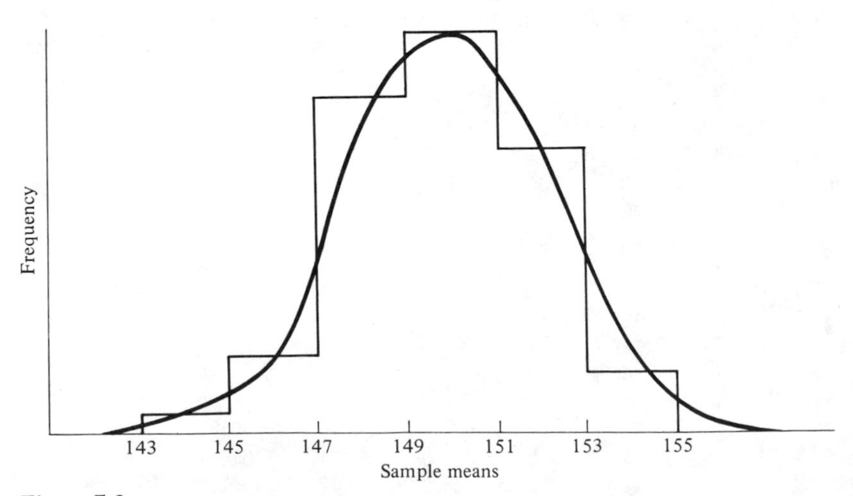

Figure 7-2

The original weights ranged over a considerable span. In the first three samples, we had one person as light as 111 pounds and someone as heavy as 180 pounds. Yet none of our sample means were anywhere near these extremes.

Although the population mean was 150 pounds, it would *not* be rare for a person to be 10 pounds over that. Notice in our first random sample that 7 persons weighed over 160 pounds. But our computer simulation indicates that it would be extremely rare to find the mean of a sample of 36 persons to be 160 pounds.

CLASS EXERCISE

This computer simulation hints at some important facts about the distribution of sample means.

7-17 Will the graph always be bell shaped? ____________

7-18 What is the theoretical average of the means? ____________

7-19 What about the range and standard deviation of the means as compared to the range and standard deviation of the original population? ____________

THE CENTRAL-LIMIT THEOREM

Statisticians have determined the following:

1. If the sample is large (and samples bigger than 30 have been found to be large enough), then the graph of the sample means will be approximately a **normal curve.**
2. Theoretically the mean of these means will be the same as the mean of the original population. In symbols, $\mu_m = \mu_{\text{pop}}$.
3. The range and standard deviation of the means will be much smaller than the range and standard deviation of the population. In fact

$$\sigma_m = \frac{\sigma_{\text{pop}}}{\sqrt{n}}$$

This can be stated formally as follows: If many, large, equal-sized random samples are drawn from a population and if the mean of each sample is computed, then the resulting distribution of sample means will be a **normal distribution.** Furthermore,

$$\mu_m = \mu_{\text{pop}} \qquad \text{and} \qquad \sigma_m = \frac{\sigma_{\text{pop}}}{\sqrt{n}}$$

In our case, we set the computer so that $\mu_{\text{pop}} = 150$ and $\sigma_{\text{pop}} = 12$. Since $n = 36$, we expect μ_m to be about 150 and σ_m to be about $12/\sqrt{36} = 12/6 = 2$. In our case, the mean of our 100 numbers came out to be 149.7 pounds and the standard deviation came out to be 2.1 pounds.

Illustration

Let us go back to the original example. We wanted to know if a sample mean of 155 should cause us to doubt that the population mean was 150 pounds.

1 Hypotheses

$H_0 : \mu_{pop} = 150$ pounds $\qquad H_a : \mu_{pop} \neq 150$ pounds

2 Justify the use of the normal distribution

Since $n = 36$ is bigger than 30, a distribution of many sample means is approximately normal. We will use $\alpha = .05$, $z_c = \pm 1.96$. We will assume that the standard deviation of the ages was 12 pounds as we had set it in the computer. In the next section we shall learn how to estimate this number from the sample data if it is unknown.

$$\sigma_m = \frac{\sigma_{pop}}{\sqrt{n}} = \frac{12}{\sqrt{36}} = 2 \qquad \mu_m = \mu_{pop} = 150$$

Thus we have

$$m_c = \mu_m + z_c \, \sigma_m$$

$$m_c = 150 + (\pm 1.96)(2) = 146.1 \text{ and } 153.9 \text{ pounds}$$

(See Figure 7-3.)

3 Decision rule

Reject H_0 if the sample mean is less than 146.1 pounds or greater than 153.9 pounds. We see by the computation that if the population mean is 150 pounds, then almost all the sample means are between 146.1 and 153.9.

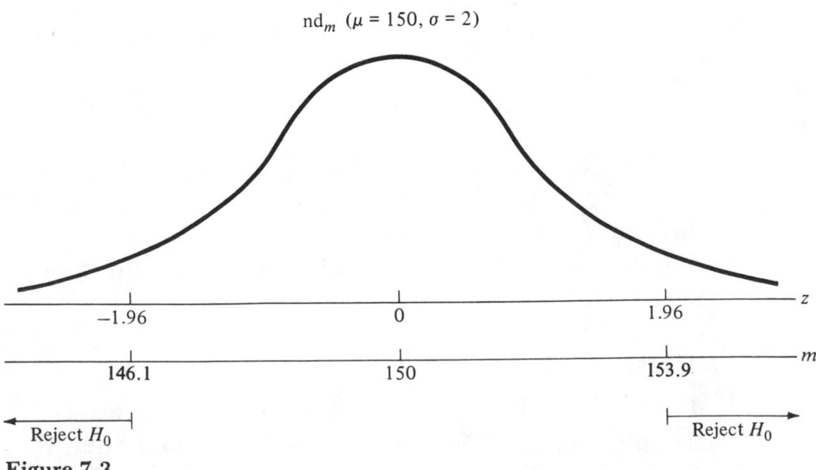

Figure 7-3

4 Outcome

The sample mean was 155 pounds.

5 Conclusion

The claim of 150 pounds as the population mean is false. The mean is more than 150 pounds.

CLASS EXERCISE

7-20 The Worstinhouse Company makes thousands of electric clothes dryers at their New Jersey factory. Their assembly line turns out dryers on which the timers last an average of 80 months with a standard deviation of 15 months. Suppose you test a random sample of 36 dryers and find out the average life of the timers. You write down the average. Then suppose you repeat this whole procedure for many days. Describe the list of numbers you will have written down. ______________________________

ESTIMATION OF THE POPULATION STANDARD DEVIATION

When the value of the population standard deviation is unknown, how can we estimate it? Statisticians symbolize this estimate as s_{pop}. We illustrate the technique with an artificial sample of just three numbers.

Suppose that a random sample of three numbers was taken from a population and the results were 8, 2, and 10. We square these numbers and get

X	X^2
8	64
2	4
10	100
$\Sigma X = 20$	$\Sigma X^2 = 168$

Adding these we get: $\Sigma X = 20$ and $\Sigma X^2 = 168$. The formula used to estimate the value of the population standard deviation is

$$s_{\text{pop}} = \sqrt{\frac{\Sigma X^2 - \frac{(\Sigma X)^2}{n}}{n - 1}}$$

This formula which estimates σ_{pop} essentially measures how far each number in the sample is from the mean of the sample and uses this information to come up with one value which is typical of this amount of variability.

In advanced statistics textbooks, it is proved that this particular formula provides an estimate for σ_{pop} which is usually close to the correct value.

In our example

$$s_{pop} = \sqrt{\frac{(168) - \frac{(20)^2}{3}}{2}} = \sqrt{\frac{168 - \frac{400}{3}}{2}} = \sqrt{\frac{168 - 133.33}{2}}$$

$$= \sqrt{\frac{34.67}{2}} = \sqrt{17.3} = 4.16$$

Estimation of the Standard Deviation of the Sample Means

If we do not know σ_{pop} we can not compute σ_m, which equals $\sigma_{pop}/\sqrt{n}$. However, if we have estimated σ_{pop} by s_{pop}, then we can estimate σ_m by $s_m = s_{pop}/\sqrt{n}$. In the example above, we found s_{pop} to be 4.16. Therefore, if we had many samples of size $n = 3$, the means of these samples would be spread out with a standard deviation of

$$s_m = \frac{s_{pop}}{\sqrt{n}} = \frac{4.16}{\sqrt{3}}$$

$$= \frac{4.16}{1.73} = 2.4$$

Our Further Example of a Hypothesis Test

Test the claim that the average weight of women's bikinis is 4 ounces if a random sample of 40 bikinis had the following weights in ounces: 2, 2.5, 3, 3, 3, 3, 3, 3.5, 3.5, 3.5, 3.5, 3.5, 3.5, 3.5, 3.5, 3.5, 4, 4, 4, 4, 4, 4, 4, 4, 4, 4, 4.5, 4.5, 4.5, 4.5, 4.5, 4.5, 4.5, 4.5, 5, 5, 5, 5, 5, 5.5. Use $\alpha = .01$.

SOLUTION

1 Define the population

The population is all weights of bikinis.

2 State the hypotheses

$H_0 : \mu_{\text{pop}} = 4$ ounces $\qquad H_a : \mu_{\text{pop}} \neq 4$ ounces

3 Compute the mean and standard deviation[*]

X	X^2	X	X^2
2	4	4	16
2.5	6.25	4	16
3	9	4	16
3	9	4	16
3	9	4	16
3	9	4	16
3	9	4.5	20.25
3.5	12.25	4.5	20.25
3.5	12.25	4.5	20.25
3.5	12.25	4.5	20.25
3.5	12.25	4.5	20.25
3.5	12.25	4.5	20.25
3.5	12.25	4.5	20.25
3.5	12.25	4.5	20.25
3.5	12.25	5	25
3.5	12.25	5	25
4	16	5	25
4	16	5	25
4	16	5	25
4	16	5.5	30.25
Totals		$\Sigma X = 157.5$	$\Sigma X^2 = 642.75$

$$s_{\text{pop}} = \sqrt{\frac{642.75 - \frac{(157.5)^2}{40}}{39}} = \sqrt{\frac{642.75 - \frac{24806.25}{40}}{39}}$$

$$= \sqrt{\frac{642.75 - 620.15625}{39}}$$

$$= \sqrt{\frac{22.59375}{39}} = \sqrt{.579} = .761$$

[*] When no confusion will arise, the estimate of the standard deviation is often referred to as the standard deviation.

4 *Justify the use of the normal curve*

$n = 40$, $40 > 30$, therefore, the distribution of sample means is approximately normal.

$$\mu_m = \mu_{\text{pop}} = 4$$

$$s_m = \frac{s_{\text{pop}}}{\sqrt{n}} = \frac{.761}{\sqrt{40}} = \frac{.761}{6.325} = .12$$

5 *Choose significance level*

$\alpha = .01 \qquad z_c = \pm 2.58$

6 *Decision rule*

(See Figure 7-4.)

$m_c = 4 + (\pm 2.58)(.12) = 4 \pm .31 = 3.69$ and 4.31

I will reject the null hypothesis if the average weight of the bikinis in my sample is less than 3.69 ounces or greater than 4.31 ounces. Our computations show that if the population average is 4 ounces, then almost all sample averages will fall between 3.69 and 4.31 ounces.

7 *Outcome*

My sample mean was $m = 157.5/40 = 3.9375 = 3.9$.

8 *Conclusion*

I fail to reject the null hypothesis. μ_{pop} might be 4 ounces.

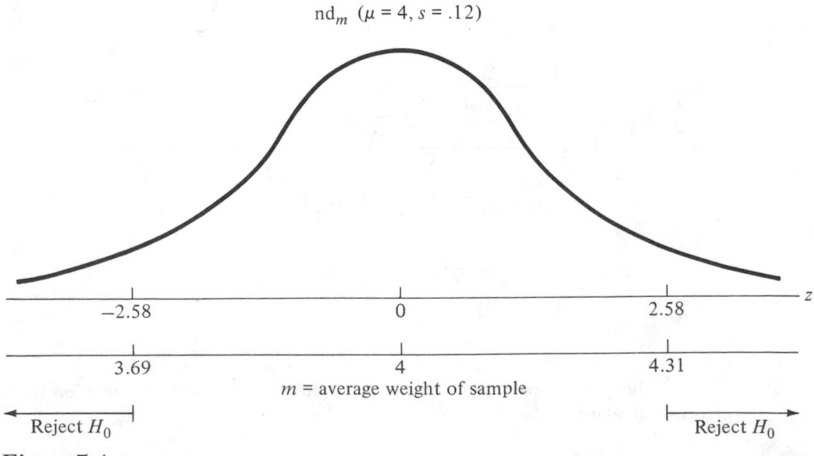

Figure 7-4

CLASS EXERCISE

7-21 Test the claim that in this book the average number of e's on a full line of text (not charts, tables, or numerical illustrations) is 5. Use $\alpha = .05$.

a Have each person in the class select one or two lines at random from the book until 36 lines have been selected.

Number of e's in your first line. ______________________

Number of e's in your second line. ______________________

b Compile all the results of the class in a frequency table, like this:

Number of e's	Frequency
0	
1	
2	
3	
4	
5	
6	
7	
8	
9	
10	
Total	$n = 36$

c Compute the *mean* of this sample. ______________________

d Compute s_{pop}. $s_{\text{pop}} =$ ______________________

e Compute s_m. $s_m =$ ______________________

f Draw a normal curve: nd_m ($\mu = 5$, $s_m =$ ________)

-1.96 1.96 z

m

g Find the two critical values of m. ________ and ________

h Do you reject the claim that the mean number of e's is 5? ____________

In Chapter 2, Assignment 2, you gathered a random sample of 100 students and found the mean of this sample. Gather the results of each student in your class. Since $n = 100$ is larger than 30, the means of all possible random samples are approximately normal. List the class outcomes here:

____

____

____

____

7-22 Graph the class's results.

7-23 Is the graph bell shaped? ______________________________

7-24 Compute the mean of these m's. ______________________________

7-25 Estimate μ_{pop}. ______________________________

7-26 Compute the standard deviation of this distribution of sample means by considering the m's as raw scores.

$$s_m = \sqrt{\frac{\Sigma m^2 - \frac{(\Sigma m)^2}{n}}{n-1}}$$

where $n =$ the number of students in this class. ______________________________

7-27 Compute s_{pop} for your individual sample.

$$s_{\text{pop}} = \sqrt{\frac{\Sigma X^2 - \frac{(\Sigma X)^2}{n}}{n-1}}$$

where $n = 100$. ______________________________

7-28 Compute

$$s_m = \frac{s_{\text{pop}}}{\sqrt{100}}$$

____

7-29 Are the answers to Exercises 7-27 and 7-29 close to each other? __________

Another Example

Dr. Legger, a biology teacher, claims that the average number of legs on the common centipede is 20. On a field trip his class gathers 81 centipedes and counts the number of legs on each. They compute m and s_{pop} from this sample of 81 centipedes.

The mean number of legs is $m = 15.7$ and $s_{pop} = 21.6$. Using $\alpha = .01$, test the doctor's claim.

1 *Define the population*

The population is the number of legs on all centipedes.

2 *Hypotheses*

$H_0 : \mu_{pop} = 20 \qquad H_a : \mu_{pop} \neq 20$

3 *Justify use of normal distribution*

$n = 81 > 30$

Therefore, we have a normal distribution of m's.

4 *Compute mean and standard deviation for the normal distribution*

$\mu_m = 20$

$$s_m = \frac{s_{pop}}{\sqrt{n}} = \frac{21.6}{\sqrt{81}} = \frac{21.6}{9} = 2.4$$

5 *Choose the significance level*

$\alpha = .01 \qquad z_c = \pm 2.58$

6 *Decision rule*

$m_c = 20 + (\pm 2.58)(2.4) = 20 \pm 6.2 = 13.8$ and 26.2

(See Figure 7-5.)

We will reject the claim that centipedes average 20 legs if our sample of 81 centipedes averages less than 13.8 legs or more than 26.2 legs.

7 *Outcome*

Our average was $m = 15.7$ legs.

8 *Conclusion*

We fail to reject the null hypothesis at this .01 significance level. Our results of 15.7 legs, although less than the 20 legs claimed by Dr. Legger, does not prove that he was mistaken.

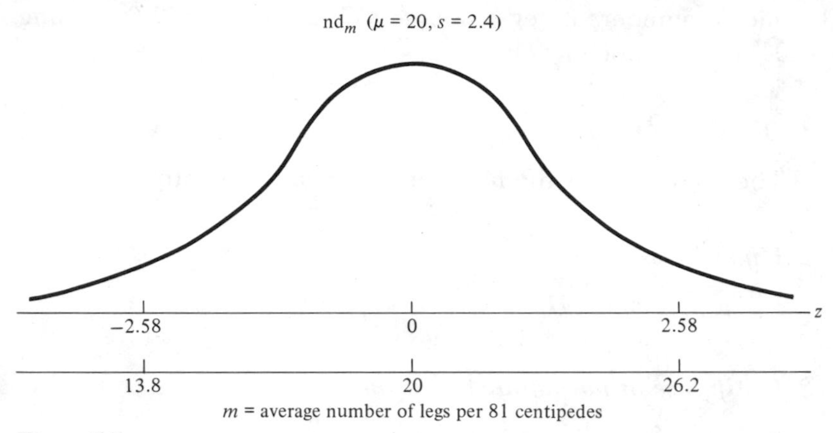

Figure 7-5

ASSIGNMENT

An average young man who was mean
To Gloria, Peggy and Jeanne
Discovered their mode
Was to quickly unload
A person who tried to demean

1. Put a penny on a sidewalk and count how many people walk past before someone picks it up. Repeat this several times and find the various averages for your results.
2. Get about 10 cubes of sugar and set each of them in a similar container of water. Record how long it takes for each one to melt completely. Compute some averages of these times.
3. Make a paper airplane. Throw it 10 times and record how far it travels each time. What is the average distance traveled?
4. Take a shuffled deck of cards. Pick five cards at random. Without looking, guess the color of each one. Record the number of times you were right. Repeat the whole experiment 10 times. Find the mean of your results.
5. Compare two brands of aspirin as to how fast they dissolve in water. Dissolve 10 pills from each brand, and find the average for each set of 10 pills.
6. Stop at a local store (preferably one with one cash register) for 5 minutes and record the total sales on the register during that time. Repeat this 10 times during the week at randomly picked times. Compute the average amount of sales during a 5-minute period. How could you use this to *estimate* total sales for the week?
7. Light 10 birthday candles. Time how long it takes for each one to burn. Compute the average time.

8 A machine which dispenses sugar is supposed to release sugar into sacks. The machine is set so that the average amount dispensed is five pounds, with a standard deviation of 0.1 pound. If an inspector finds that a sample of 100 sacks has a mean weight of 4.9 pounds, should she assume that something is wrong with the machine? Use $\alpha = .01$.

9 It is known that scores on a certain music aptitude test have mean 100 and $\sigma = 10$ when given to adults trained in American music schools. The test was given to 64 adults trained in British music schools and their average score was 105. Does this indicate at the .05 significance level that the mean on this test for British students is different from the mean for Americans?

10 It is known from long experience that the average number of mail orders received each day by a large mail order company is 1,100 with $\sigma = 50$. The company is going to start a new advertising policy. It is then going to check the mail for 100 days to find the new average. What results are necessary to decide if the policy has changed the average number of orders per day? Use $\alpha = .05$.

11 a Here are the ages of the people who are going to be at a party: 6, 16, 22, 47, 47, 66. Compute the mean age of people at this party.

b Here are the ages of the people attending another party: 30, 35, 33, 37, 37, 34, 32. Compute the mean age of people at this party.

c In which example, a or b, is the average age more representative of the ages of the people at the party. Complete these sentences: When the numbers in a list are close to the average, then the average is (more)(less) representative of the list. When the numbers in a list are scattered far away from the average, then the average is (more)(less) representative of the list.

In Assignments 12, 13, and 14, compute m, s_{pop}, and s_m from the following sample data. What does each of these values measure?

12 Number of children in a family: 5, 0, 2, 0, 1, 10, 3, 3, 2, 4, 3, 2, 1, 1, 1, 0, 17, 2, 2, 3, 0, 4, 1, 8, 5, 2, 3, 4, 2, 6, 0, 11, 2, 3, 1, 8.

13 Weights of children: 40, 32, 60, 80, 45, 40, 80, 71, 20, 16, 42, 37, 80, 75, 69, 80.

14 Grade point averages of ΩΔΣ sorority: 3.5, 3.0, 3.1, 3.4, 2.5, 2.9, 3.8, 3.7, 3.0, 3.0, 3.6, 3.8, 3.6, 3.5, 3.1, 2.8, 3.1, 3.3, 3.8, 3.0, 3.1, 3.4, 3.5, 3.4, 3.3

15 Which of the above sample means is a member of a normal distribution of sample means?

16 Dr. P. Ramitter ran the ages of all the students on his campus through a computer and determined that $\mu_{\text{pop}} = 19.8$ years and $\sigma_{\text{pop}} = 4.8$ years. He assigns his students the task of gathering a random sample of 64 students' ages and of computing m.

a 95 percent of the students who do the assignment correctly should find m between __________ and __________.

b *True or False* If he decides to fail all the students who turn in a sample mean outside this range, he runs a .05 chance of failing a student who should pass.

c *True or False* If he passes all the students who are inside this range, he runs a .95 chance of passing a student who did not gather a random sample.

17 Using the .01 significance level, test the claim that the average height of American GIs is 5 feet 10 inches if a sample of 400 soldiers had a mean height of 5 feet 10.8 inches with $s_{pop} = 4.2$ inches.

18 At a community meeting, the manager of Garden Plaza stated that the average age of adults in his building was 32 years. A sample of 40 adults had an average of 29.1 years with a standard deviation of 6.4 years. Do these figures support or discredit the manager's claim? Use $\alpha = .05$.

19 The mean cumulative average of all the girls in Sigma Epsilon Chi sorority is 2.91. A random sample of 100 other girls at County College had a mean cumulative average of 2.86 with $s_{pop} = 3.0$. Perform a hypothesis test at the .05 significance level using these figures.

20 A newspaper stated that the average age of Eagle Scouts was 15 years old. A local organization of 38 Eagle Scouts had an average age of 13.3 years old with a standard deviation of 4.2 years. Test at the .01 significance level. What exactly are you testing?

21 A claim is made in *T.V. Guide* that the average American family watches television $5\frac{1}{2}$ hours per day. In a sample of 50 families in Suffolk county, the average was 4.2 hours with $s_{pop} = 1.8$ hours. Using $\alpha = .05$, perform a hypothesis test. State clearly what you are testing.

22 The chief of security at Mobile University claims that the average car on campus carries $1\frac{3}{4}$ persons. In a random sample of 300 cars at Mobile U., we averaged 1.5 persons with $s_{pop} = .65$. Test with $\alpha = .01$.

23 A medical textbook by Dr. Arthur Writus claims that the average pulse rate for an adult male is 80 beats per minute. A random sample of 40 adult male patients at a local hospital had the following pulse rates: 56, 52, 88, 100, 82, 92, 76, 104, 92, 56, 80, 92, 86, 90, 64, 75, 89, 94, 102, 58, 62, 88, 52, 76, 91, 84, 89, 63, 58, 62, 101, 92, 98, 83, 75, 84, 91, 86, 72 and 92. Test using $\alpha = .05$. Is the population being sampled all adult males? What are you testing?

24 Percy Kushun wrote in *Drumbeat* magazine that the average age at which a person learns to play a musical instrument is $15\frac{1}{2}$ years. A random sample of customers in a local music shop uncovered the following data: 20, 8, 23, 7, 17, 18, 9, 8, 6, 12, 35, 50, 8, 8, 9, 17, 10, 8, 12, 42, 8, 29, 30, 9, 10, 87, 12, 18, 20, 13, 9, 5, 10, 19, 20, 25. Test using $\alpha = .05$. What is the population being tested? What are you testing?

SMALL-SAMPLE TESTING

Sometimes an experimenter cannot control the size of the sample. For example, if a teacher wishes to use his class as a sample and he has 23 students, then $n = 23$ is not greater than 30, and the distribution of sample means cannot be assumed to be normal. You have seen that as you must work with smaller and smaller samples, it gets harder and harder to come to accurate decisions about the population. In general, it is much easier to test hypotheses when you have large samples. The effects of random sampling are more consistent when you have large samples. For example, you can see how difficult it would be to predict the average amount of money men carry in their wallets by asking just two men. If you sample only two people, you might just get two extreme cases. However, if you sample 200 people, it is not likely that you will get 200 extreme cases.

The theory of small-sample statistics is, therefore, more complicated than the theory of large-sample statistics. In fact, if we take many small samples from a population and construct a distribution of sample means, it is not simple to predict what pattern they will form. However, it can be shown mathematically that *if the original population is itself normally distributed,* then even with small samples, the sample means fall into a predictable pattern. Because many populations are approximately normally distributed, we will go ahead to discuss small-sample statistics from normal populations. For approaches to problems where we cannot assume that the populations are normal, see chapter 11.

On the first few pages of this chapter, we discussed what would happen if we gathered a random sample of 36 weights from a distribution of weights with $\mu_{\text{pop}} = 150$ pounds and $\sigma_{\text{pop}} = 12$ pounds. We saw that the mean of such a sample would be near 150 pounds and that, furthermore, *if* $n > 30$, then a distribution of such means would be approximately normal and that, therefore, about 95 percent of these sample means would be within 1.96 standard deviations of the mean. In the case where n equalled 36, we found that 95 percent of the sample means should lie between 146.08 and 153.92 pounds.

However, smaller samples give less accurate estimations for a population. The mean of only 16 weights might well be further from 150 pounds than the mean of 36 weights. There will be more variation in a distribution of means based on small samples. We might find more than 5 percent of the sample means outside the range from 146.08 to 153.92 pounds.

COMPUTER SIMULATION

We programmed a computer to pick a random sample of 16 weights from the population of weights with $\mu = 150$ pounds and $\sigma = 12$ pounds. The population was normally distributed to begin with. Here are the results:

152.773	140.07	184.696	176.72	165.459	139.873	158.784	160.271
128.494	125.784	136.22	142.12	138.641	128.14	156.011	163.923

The mean of this sample is 149.936.

We had the computer repeat this procedure, and we got:

135.726 157.623 163.447 162.252 176.596 169.499 147.745
158.193 152.975 174.651 148.457 138.388 166.694 149.557
151.845 127.12

The mean of this sample is 155.048.

A third computer run produced the following:

161.069 130.709 153.619 178.383 159.775 155.994 166.202
146.421 122.369 159.496 156.661 149.051 147.96 160.847
145.83 174.523

The mean of this sample is 154.307.

TABLE 7-3

Numerical Order	*m*	Numerical Order	*m*	Numerical Order	*m*
1	142.698	35	148.58	68	151.024
2	143.223	36	148.586	69	151.043
3	143.76	37	148.607	70	151.142
4	143.834	38	148.794	71	151.2
5	144.617	39	148.886	72	151.345
6	144.624	40	148.913	73	151.47
7	144.74	41	149.007	74	151.641
8	144.946	42	149.145	75	151.962
9	145.687	43	149.362	76	152.134
10	145.734	44	149.364	77	152.147
11	145.782	45	149.412	78	152.257
12	146.002	46	149.498	79	152.4
13	146.298	47	149.584	80	152.534
14	146.33	48	149.669	81	152.854
15	146.419	49	149.776	82	152.884
16	146.567	50	149.936	83	152.962
17	146.64	51	149.988	84	152.993
18	146.731	52	150.1	85	152.998
19	146.747	53	150.1	86	153.027
20	146.819	54	150.233	87	153.007
21	146.901	55	150.352	88	153.468
22	146.937	56	150.363	89	153.656
23	146.943	57	150.391	90	153.861
24	146.986	58	150.442	91	154.269
25	147.242	59	150.516	92	154.277
26	147.419	60	150.533	93	154.307
27	147.587	61	150.6	94	154.354
28	147.663	62	150.652	95	154.526
29	147.896	63	150.666	96	154.733
30	147.904	64	150.756	97	155.012
31	147.925	65	150.78	98	155.048
32	148.002	66	150.817	99	155.456
33	148.006	67	150.924	100	156.067
34	148.328				

TABLE 7-4

Interval	Frequency
142–144	4
144–146	7
146–148	20
148–150	20
150–152	24
152–154	15
154–156	9
156–158	1

With large samples, our first 3 sample means were 151.8, 148.5, and 149.6 pounds. Now with a small sample, we got 149.9, 155.0, and 154.3. As we expected, our sample means are more varied. We had the computer run this program 97 more times and then arrange the sample means in numerical order. The printout is shown in Table 7-3. You will notice that 12 out of 100 or 12 percent are less than 146.08 pounds, and 10 out of 100 or 10 percent are greater than 153.92 pounds.

A graph of this distribution will show that this distribution of sample means is still a bell-shaped curve. However, it can be shown that it is not quite the shape of a normal curve.

A frequency chart for the sample means is given in Table 7-4. The graph is given by Figure 7-6.

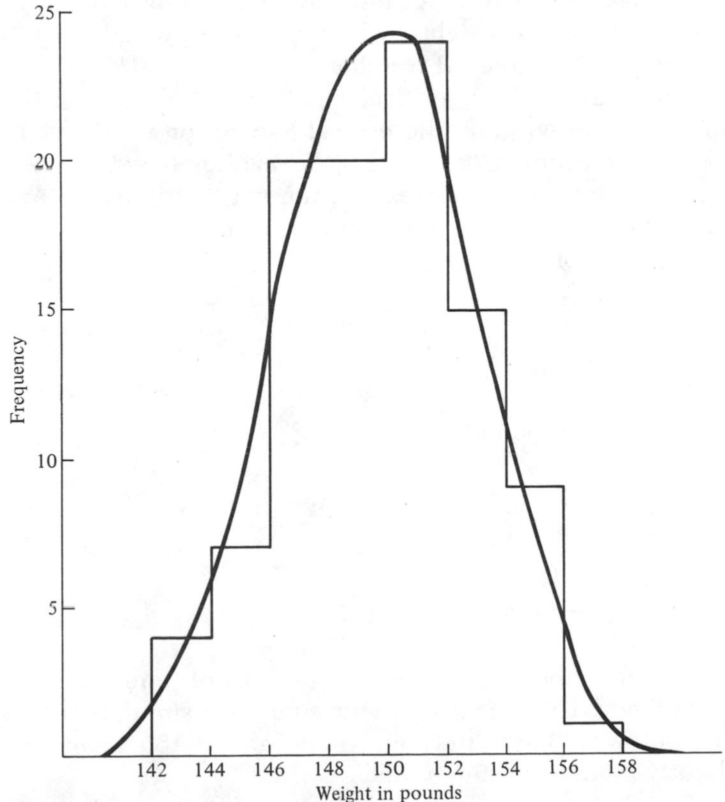

Figure 7-6

Under the assumption that the original population was approximately a normal distribution, the bell-shaped curves arising from small samples are called **Student's *t* curves.** They are so named because the first person to write about them, Mr. W. Gossett, used the pen name A. Student. A set of measurements whose graph is a t curve is said to have a **t distribution.** We shall use the abbreviation td for t distribution. A hypothesis test which uses a t curve is called a **t test.**

In our computer simulation, we set up the original population so that it was normally distributed. We assume throughout this chapter that the samples are drawn from populations which are approximately normal. Using more advanced methods, it is possible to test a population for normalcy by using a chi-square test.

There are many t curves, each a differently shaped bell curve. The exact shape depends on the size of the samples used to create the distribution. The various t curves are classified in tables according to their degrees of freedom. For a one-sample hypothesis test involving sample means, the correct value for the degrees of freedom is $n - 1$, where n is the sample size. This is related to the fact that in picking n numbers to have a certain mean, you are free to pick $n - 1$ of them to be anything at all. For example, in picking 16 weights whose mean is equal to 150 pounds, there are 15 degrees of freedom.

If you turn to the table of critical t scores in the back of the book, Table B-4, you will find the critical t score for $\alpha = .05$ at 15 degrees of freedom equal to ± 2.13. We can interpret these t scores just as we interpret z scores in a normal distribution. We expect about 95 percent of our sample means to be within ± 2.13 standard deviations of the mean, that is,

$$m_c = \mu_m + t_c \sigma_m$$

where $\mu_m = \mu_{\text{pop}} = 150$ pounds

$$\sigma_m = \frac{\sigma_{\text{pop}}}{\sqrt{n}} = \frac{12}{\sqrt{16}} = \frac{12}{4} = 3$$

Thus $m_c = 150 + (\pm 2.13)(3) = 150 \pm 6.4$

Thus, we expect about 95 percent of our sample means to lie between 143.6 and 156.4 pounds. Our computer simulation had two means below 143.6 pounds and none above 156.4 pounds. This is slightly more than 95 percent.

Example of a *t* test

Note, in the following example, that the only changes from the large-sample procedure are the using of a critical t score instead of a z score from the normal distribution and the assumption that the population is normal.

A sociology report states that, on the average, college students intend to have two children. To test whether this is true at her school, Valentine Hubbard surveys her English lit. class asking how many children each student intends to have. Here are the 25 replies: 3, 0, 0, 1, 2, 2, 0, 4, 3, 4, 2, 2, 1, 0, 1, 0, 1, 1, 0, 2, 3, 5, 10, 2, 0. Let $\alpha = .05$.

1 Identify the population

Numbers of children that students at Valentine's college intend to have.

2 State the hypotheses

$H_0 : \mu_{\text{pop}} = 2 \qquad H_a : \mu_{\text{pop}} \neq 2$

3 Which distribution?

$n = 25 \not> 30$

therefore, a t distribution will be used.

$\text{df} = n - 1 = 24$

4 Compute the mean and the standard deviation

X	X^2	X	X^2
0	0	2	4
0	0	2	4
0	0	2	4
0	0	2	4
0	0	2	4
0	0	3	9
0	0	3	9
1	1	3	9
1	1	4	16
1	1	4	16
1	1	5	25
1	1	10	100
2	4		
		$\Sigma X = 49$	$\Sigma X^2 = 213$

We get

$$s_{pop} = \sqrt{\frac{213 - \frac{(49)^2}{25}}{24}} = \sqrt{\frac{213 - \frac{2401}{25}}{24}} = \sqrt{\frac{213 - 96.04}{24}}$$

$$= \sqrt{4.873} = 2.208$$

Therefore,

$$s_m = \frac{2.208}{\sqrt{25}} = \frac{2.208}{5} = .4416$$

So, we have

$td_m\ (\mu = 2,\ s = .4416)$

5 *Significance level*

$\alpha = .05$

Looking up 24 df with $\alpha = .05$ in Table B4, we find $t_c = \pm 2.06$.

6 *Decision rule*

$m_c = 2 + (\pm 2.06)(.4416) = 2 \pm .91 = 1.09$ and 2.91 (See Figure 7-7.)

I will reject the null hypothesis if the mean number of children intended in my sample is less than 1.09 or greater than 2.91.

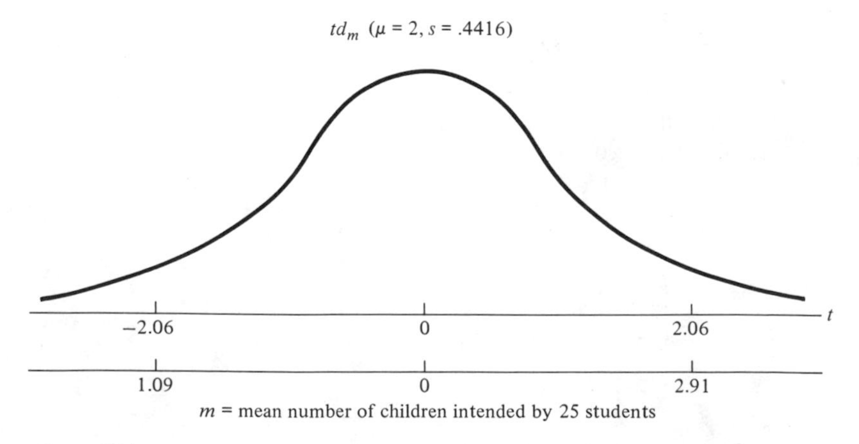

Figure 7-7

7 Outcome

My sample mean was $m = 49/25 = 1.96$.

8 Conclusion

I fail to reject the null hypothesis. The report figure of two children may well be correct based on this school's population.

Another Example

Test the hypothesis that the mean salary of grade school teachers in Eduville is \$11,000. Use $\alpha = .01$. A random sample of 16 teachers has a mean of \$12,000 and $s_{\text{pop}} = \$100$.

1 Identify population

Salaries of Eduville grade school teachers.

2 State the hypotheses

$H_0 : \mu_{\text{pop}} = \$11{,}000 \qquad H_a : \mu_{\text{pop}} \neq \$11{,}000$

3 Which distribution?

$n = 16 \not> 30$, therefore, a t distribution.

4 Find mean and standard deviation

$\mu_m = \mu_{\text{pop}} = \$11{,}000$

$$s_m = \frac{s_{\text{pop}}}{\sqrt{n}} = \frac{100}{\sqrt{16}} = \$25$$

So we have td_m ($\mu = \$11{,}000$, $s = \$25$).

5 Significance level

$\alpha = .01 \qquad \text{df} = 15 \qquad t_c = \pm 2.95$

6 Decision rule

(See Figure 7-8.)

$m_c = \$11{,}000 + (\pm 2.95)(\$25) = \$11{,}000 \pm \$73.75 = \$10{,}926.25$ and \$11,073.75

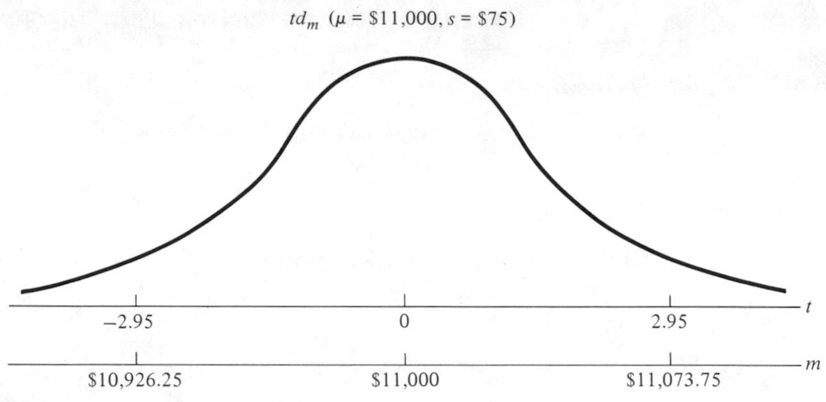

Figure 7-8

I will reject the null hypothesis if the average salary in my sample is less than $10,926.25 or greater than $11,073.75.

7 *Outcome*

The mean of my sample was $12,000.

8 *Conclusion*

I reject the null hypothesis. I am more than 99 percent certain that the mean salary is larger than $11,000.

Chapter Summary

We have seen that there are several types of useful averages. Of these, we are particularly interested in the sample mean because the distribution of sample means in repeated sampling is often known. The central-limit theorem states that sample means of *large* samples from any population are normally distributed. The work of W. Gossett showed that means of *small* samples from *normal* populations have the so-called t distribution. Knowing these distributions allows us to judge the likelihood that a particular sample mean has come from any specified population.

Words and Symbols You Should Know

Distribution of sample means	Mode	m	td
Central-limit theorem	Sample mean	m_c	μ_m
Mean	Small-sample testing	s_m	μ_{pop}
Median	Student's t distribution	s_{pop}	σ_m
Midrange		t	σ_{pop}
		t_c	

TABLE 7-5 SUMMARY OF HYPOTHESIS TESTS STUDIED THUS FAR

Name of Test	When Used	Conditions	Formulas
I Chi Square	Are two variables related?	Each $E > 5$ Not 2×2 Use chi-square distribution	1 $E = \frac{(RT)(CT)}{n}$ 2 $X^2 = \Sigma \frac{O^2}{E} - n$ 3 a $\text{df} = C - 1$ (if $R = 1$) b $\text{df} = (R-1)(C-1)$ (if $R \neq 1$)
II Binomial One-Sample	Is a claim about a proportion correct?	$n\pi$ and $n(1-\pi) > 5$ Use nd	1 $\mu_p = \pi$ 2 $s_p = \sqrt{\frac{\pi(1-\pi)}{n}}$
III Binomial Two-Sample	Are two proportions equal?	Four outcomes all greater than 5 Use nd	1 $\mu_{dp} = \pi_1 - \pi_2$ 2 $s_{dp} = \sqrt{\frac{pq}{n_1} + \frac{pq}{n_2}}$ 3 Outcome $= p_1 - p_2$
IV A Sample-Mean, One Large Sample	Is a claim about a mean correct?	$n > 30$ Use nd	1 $\mu_m = \mu_{\text{pop}}$ 2 $s_{\text{pop}} = \sqrt{\frac{\Sigma X^2 - \frac{(\Sigma X)^2}{n}}{n-1}}$ 3 $s_m = \frac{s_{\text{pop}}}{\sqrt{n}}$
IV B Sample-Mean, One Small Sample	Is a claim about a mean correct?	$n \leq 30$ and population is normal Use *td*	1 $\text{df} = n - 1$

ASSIGNMENT

A. Student viewed lads who were dirty
He sampled a few less than 30
Instead of score z
He tested with t
And his results came out purty

What critical t score or critical z score would you use in the following one-sample hypothesis tests of sample means?

	n	α	t_c or z_c
25	17	.01	
26	35	.05	
27	11	.01	
28	30	.05	

29 Identify the following as:

A Chi square
B Binomial one-sample
C Binomial two-sample
D Sample-mean, large sample
E Sample-mean, small sample
F Other

a 30 percent of male doctors smoke.
b The average height of MPs is 6 feet 6 inches, given that the average height of 100 MPs is 6 feet 5 inches.
c Number of children and religion are independent.
d The average weight of Wacs is the same as the average weight of Waves.
e The percentage of Marines who do is the same as the percentage of Coast Guard who do.
f The proportion of fishermen who lie is 80 percent, given that 43 lied in a sample of 50.
g Rank in senior-year high school class can predict rank in freshman-year college class.
h The average weight of dinosaur eggs is 3.5 pounds, given that eight eggs averaged 2.8 pounds.

30 A hospital management spokesman states that the average hospital employee earns $5.50 an hour. A newspaper reporter samples eight striking staff members. Their mean salary is $5.07 an hour with $s_{\text{pop}} = \$1.71$. Test the spokesman's claim at the .05 significance level.*

31 The BIM computer monthly bulletin states that the average age of employees at BIM is 32 years. A random sample of 20 employees in the lounge area had an average age of 30.8 years with a standard deviation of 1.4 years. Does this evidence tend to support or deny the claim in the bulletin? Use $\alpha = .01$.

32 A magazine advertisement states that children average 11 cavities between the ages of 5 and 15 years. A random sample of 10 15-year-old children's dental records gave the following data:

$$m = 12.1 \qquad s_{\text{pop}} = 4.53$$

Test with $\alpha = .01$.

* In all problems in this book which deal with small samples, assume that the populations sampled are normal.

33 A report states that the average retirement age of teachers is 80. A random sample of nine teachers has an average retirement age of 70 with a standard deviation equal to 12. Do a hypothesis test at the .01 significance level.

34 It is claimed that gray squirrels have, on the average, 12 fleas. Yesterday 100 gray squirrels were inspected and the number of fleas on each was counted. m was computed to be 15 and $s_{\text{pop}} = 16$. Perform a hypothesis test with $\alpha = .05$.

35 *Local Geographic* magazine states that the average weight of hippos is 3,225 pounds. A sample of 12 hippos yields an average weight of 3,500 pounds with $s_{\text{pop}} = 500$ pounds. Test with $\alpha = .05$.

36 A spokesman for the barbers' union states that men get a haircut about every 5 weeks. A survey of 25 men showed that the average length of time was 4.2 weeks and that the standard deviation was 1.0 weeks. Does this evidence tend to support or deny the union spokesman's claim? Use $\alpha = .05$.

37 The average PIQ is supposed to be 100. Dr. Iva Pet claims to have devised a way to test the PIQ of porpoises. The average PIQ of 50 porpoises was found to be 105 with $s_{\text{pop}} = 14$. Test at the .01 significance level.

38 Test the claim that the average score on a test which measures a person's tendency to conformity is 12.3 if eight persons scored: −3, 0, 5, 8, 12, 19, 25, and 28. Use $\alpha = .05$.

39 A college newspaper states that women at that school eat 1.9 meals per day. Test that claim at the .01 significance level, using the sample data below:

  9 women ate 1 meal
  10 women ate 2 meals
  8 women ate 3 meals
  2 women ate 4 meals

40 a Secure a claim about an average from a newspaper, a magazine, a textbook, radio, television, or a teacher.
  b Outline a hypothesis test about this number. Be clear about the population to be tested, the sample to be gathered, etc.
  c Carry out the experiment which you have outlined in part b.

41 Make up your own question and use the data in Appendix A to answer it.

Readings from J. Tanur (ed.), "Statistics: A Guide to the Unknown," Holden-Day, San Francisco, 1972.

42 Setting Dosage Levels, Dixon, p. 34. How to make a good cocktail. What is sequential testing?

43 Police Manpower Versus Crime, Press, p. 112. How can you measure the "average" amount of crime in a district? How do problems of reporting crime affect such a study?

44 Varieties of Military Leadership, Selven, p. 253. Using average ratings to describe officers.

Review

45 A student designed a project to see if there was any relationship between a student's knowledge of the meaning of impeachment and his attitude toward im-

peachment. She gathered the following data by questioning a random sample of fellow students. She asked them the following questions.

1 If the President were impeached today, who would be President tomorrow?
  a The President
  b The Vice President
  c Neither the President nor the Vice President
2 Do you think that the President should be impeached?

The data she obtained:

Answer to Question 1	In Favor of Impeachment	Against Impeachment	Undecided
correct	70	30	0
wrong	29	150	21

Perform a hypothesis test using $\alpha = .01$. What are the null and alternative hypotheses? How can you interpret the results of the test?

46 The Friendship Ice Cream Company directs its managers that all orders of ice cream are to include at least 40 percent of vanilla. George Van Dyne, a manager, believes that this is wrong for his locale. In a random sample of 200 ice cream cones ordered, 70 were vanilla. Does this support his claim at the .05 significance level?

47 A student surveyed two groups of women on campus asking them if they preferred guys to wear jeans or dress pants on campus. When asking the first group, he wore jeans. When asking the second group, he wore dress pants. 22 out of 90 girls in the first group preferred dress pants. 21 out of 100 in the second group preferred dress pants. Using $\alpha = .01$, is this sufficient evidence to show that the attire worn by the interviewer affected the responses given?

EIGHT

Comparing Averages

Strange all this difference should be
'twixt tweedle dum and tweedle dee.

John Byron

Do Aquarians and Capricorns spend the same amount of sleep time engaged in *dreaming?* We would have to answer such a question by comparing the *average* dreaming time for Aquarians with the *average* dreaming time for Capricorns. This type of comparison is carried out in what are called **two-sample sample-mean tests.**

Suppose we took a random sample of 50 Aquarians and found that the average dream time was 102 minutes per night. We write $m_1 = 102$. Then we gathered a random sample of 32 Capricorns and found their average dream time to be 90 minutes. We write $m_2 = 90$. The difference between m_1 and m_2 is 12 minutes. We have to judge whether or not this difference in sample statistics is large enough to reflect a real difference in the population parameters. If the mean dream time for Aquarians and Capricorns is really the same, would we be likely to get a sample difference of 12 minutes?

The answer depends on two things. First of all, it depends on the *sample sizes.* Obviously, large samples usually give a more reliable picture. Secondly, it depends on the *variability* of each population. If the populations have large standard deviations, then the sample results will vary a lot, and a big sample difference will not be a sure indication of a population difference.

CLASS EXERCISE

8-1 For a term paper, Stanley is comparing the musical sensitivity of Leos and Virgos on his campus. Because of other pressing business, he was unable to start on the paper earlier and it is due tomorrow. He gets a random sample of two Leos and, for comparison, a random sample of two Virgos. He gives each person a musical sensitivity test, and he finds that the two Leos have scores of 89 and 91 and the two Virgos have scores of 139 and 141. He concludes that the average score for Virgos is about 50 points higher than the average score for Leos. Criticize his work as specifically as you can. In particular, what would you expect to gain by choosing bigger samples? ______________________

COMPUTER SIMULATION

We asked before if a sample difference of 12 minutes was significant. In order to get some idea of whether a difference of 12 minutes between the means of two samples is significant or not we devised a computer simulation. We created two populations. The first one had a true average dream time of 100 minutes and a standard deviation of 10 minutes. The second one had a true mean of 100 minutes and a standard deviation of 8 minutes. We then picked a

TABLE 8-1

Result Number	m_1	m_2	$dm = m_1 - m_2$
1	102.01	100.15	1.86
2	100.30	99.63	0.67
3	100.18	99.97	0.21
4	98.78	102.40	−3.62
5	97.20	99.72	−2.52

random sample of 50 numbers from the first population. The mean of this sample was 102.01. We then had the computer pick a random sample of 32 numbers from the second population. The mean of the second sample was 100.15. If we symbolize the *d*ifference of the *m*eans by *dm*, then

$$dm = m_1 - m_2 = 102.01 - 100.15 = 1.86$$

This is certainly much smaller than 12. But to make a pattern more clear, we had the computer repeat this 4 more times. The results are in Table 8-1. Finally we had the computer go through this procedure 100 times and compute 100 sample differences. The printout of all 100 results are arranged in numerical order in Table 8-2.

Looking at this set of results, we can see clearly that a sample difference of 12 minutes is much bigger than anything appearing in this list. We can be very sure that if our random samples of Aquarians and Capricorns show a 12-minute difference, then there really is a difference in the average dream time for the two groups. Such a big difference is *not* due to the variation introduced in the sampling procedure.

CLASS EXERCISE

8-2 We knew that $\mu_1 = 100$ and $\mu_2 = 100$. Therefore, we expected each value of m_1 to be near ________ and each value of m_2 to be near ________.

8-3 $dm = m_1 - m_2$. We expected each of the 100 values of dm to be near

____.

8-4 If we average these 100 dm's, theoretically we expect the mean to equal ________. We write $\mu_{dm} = \mu_1 - \mu_2$.

In our computer simulations the mean of our 100 outcomes was −.4. The graph of the dm's can be drawn by first grouping the 100 outcomes into a frequency chart as shown in Table 8-3. The graph of these results appears in Figure 8-1. Smoothing out this

TABLE 8-2

Result Number	$dm = m_1 - m_2$	Result Number	$dm = m_1 - m_2$	Result Number	$dm = m_1 - m_2$
1	−5.22	35	−1.07	68	0.21
2	−4.51	36	−1.06	69	0.32
3	−3.91	37	−1.03	70	0.42
4	−3.89	38	−1.02	71	0.46
5	−3.66	39	−0.88	72	0.60
6	−3.62	40	−0.83	73	0.65
7	−3.58	41	−0.83	74	0.67
8	−3.51	42	−0.82	75	0.80
9	−3.02	43	−0.79	76	0.84
10	−2.99	44	−0.78	77	0.86
11	−2.96	45	−0.66	78	0.90
12	−2.95	46	−0.61	79	0.92
13	−2.63	47	−0.59	80	0.95
14	−2.52	48	−0.58	81	1.04
15	−2.52	49	−0.46	82	1.05
16	−2.44	50	−0.45	83	1.09
17	−2.38	51	−0.43	84	1.19
18	−2.36	52	−0.38	85	1.29
19	−2.27	53	−0.37	86	1.57
20	−2.14	54	−0.35	87	1.57
21	−1.79	55	−0.26	88	1.70
22	−1.76	56	−0.18	89	1.70
23	−1.69	57	−0.14	90	1.70
24	−1.62	58	−0.13	91	1.86
25	−1.49	59	−0.08	92	1.86
26	−1.39	60	0.03	93	1.94
27	−1.36	61	0.05	94	2.90
28	−1.35	62	0.07	95	2.97
29	−1.26	63	0.08	96	3.08
30	−1.19	64	0.12	97	3.20
31	−1.18	65	0.12	98	3.25
32	−1.17	66	0.20	99	3.58
33	−1.17	67	0.21	100	5.75
34	−1.09				

TABLE 8-3

Interval	Frequency	Interval	Frequency
−6 to −5	1	0 to 1	21
−5 to −4	1	1 to 2	13
−4 to −3	7	2 to 3	2
−3 to −2	11	3 to 4	4
−2 to −1	18	4 to 5	1
−1 to 0	21		

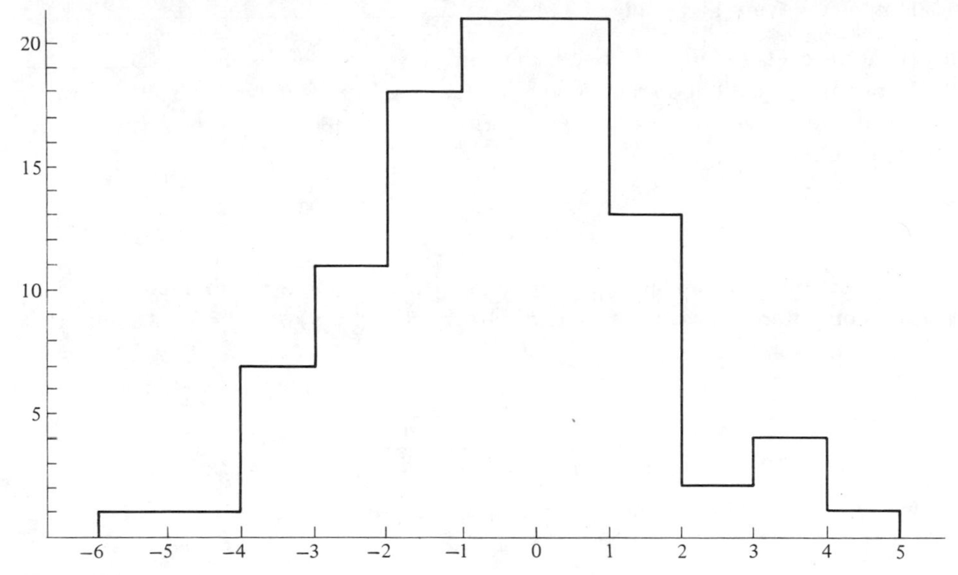

Figure 8-1

graph leads to a bell-shaped curve, and in general if n_1 and n_2 are both greater than 30, then it will be approximately a normal curve.

STANDARD DEVIATION OF THE DIFFERENCES

Statisticians have shown that the standard deviation of the *dm*'s can be found by the formula

$$\sigma_{dm} = \sqrt{\frac{\sigma_1^2}{n_1} + \frac{\sigma_2^2}{n_2}}$$

Since we had set up the computer populations ourselves, we know σ_1 and σ_2. If you recall, we made $\sigma_1 = 10$ and $\sigma_2 = 8$.

CLASS EXERCISE

8-5 Compute the theoretical value of σ_{dm} using the formula

$$\sigma_{dm} = \sqrt{\frac{\sigma_1^2}{n_1} + \frac{\sigma_2^2}{n_2}}$$

$\sigma_{dm} =$ ____________________

8-6 Since $\sigma_m = 2$ and $\mu_{dm} = \mu_1 - \mu_2 = 100 - 100 = 0$
we have nd_{dm} $(\mu = 0,\ \sigma = 2)$. We expect about 95 percent of our *dm*'s to fall between $0 + (\pm 1.96)(2) = 0 \pm 3.92$.

How many of our *dm*'s were greater than 3.92? ____________________

8-7 How many were less than −3.92? ____________________

8-8 What percentage fell within 1.96 standard deviations of the mean? __________

8-9 In practice, we often do not know σ_1 and σ_2. And, in fact, we have only *one* pair of samples. We then estimate σ_1 and σ_2 by s_1 and s_2. The estimate for the value of σ_{dm} is given by

$$s_{dm} = \sqrt{\frac{s_1^2}{n_1} + \frac{s_2^2}{n_2}}$$

For example, in the computer simulation, the first time we took 50 Aquarians and computed the mean of their 50 dream times we got $m_1 = 102.01$ minutes. Using the formula

$$s_1 = \sqrt{\frac{\Sigma X^2 - \frac{(\Sigma X)^2}{50}}{49}}$$

we found $s_1 = 11.02$ minutes. Similarly, for the 32 dream times of the Capricorns, we found $m_2 = 100.15$ and

$$s_2 = \sqrt{\frac{\Sigma X^2 - \frac{(\Sigma X)^2}{32}}{31}} = 8.91$$

Use these values of s_1 and s_2 to estimate σ_{dm}. ____________________

Summarizing the above we have: Two large random samples (each bigger than 30) are taken from two populations and the two sample means are computed. The difference between these sample means, *dm*, is found. If this procedure is repeated over and over again,

1 The distribution of *dm*'s will be normal.

2 $\mu_{dm} = \mu_1 - \mu_2$

3 $\sigma_{dm} = \sqrt{\frac{\sigma_1^2}{n_1} + \frac{\sigma_2^2}{n_2}}$ and $s_{dm} = \sqrt{\frac{s_1^2}{n_1} + \frac{s_2^2}{n_2}}$

Illustration

We return to our original question: Is the average dream time for Aquarians equal to the average dream time for Capricorns?

1 State the populations

pop_1 = Aquarians pop_2 = Capricorns

2 *Write the hypothesis*

H_0: There is no difference between μ_1, the average dream time for Aquarians, and μ_2, the average dream time for Capricorns.

$$\mu_{dm} = \mu_1 - \mu_2 = 0$$

$$H_a : \mu_{dm} \neq 0$$

3 *Which distribution?*

Since both $n_1 = 50$ and $n_2 = 32$ are greater than 30, we have an approximately normal distribution of differences.

4 *Compute means and standard deviations*

Suppose for the 50 Aquarians we compute $m_1 = 102$ and $s_1 = 11$. For the 32 Capricorns we get $m_2 = 90$ and $s_2 = 10$. Then

$$s_{dm} = \sqrt{\tfrac{121}{50} + \tfrac{100}{32}} = \sqrt{5.545} = 2.35$$

Since our null hypothesis assumes that $\mu_{dm} = 0$, we get nd_{dm} $(\mu = 0,\ s = 2.35)$.

5 *Decision rule*

If $\alpha = .01$, then $z_c = \pm 2.58$. Therefore,

$$dm_c = 0 + (\pm 2.58)(2.36) = \pm 6.1$$

I will reject the null hypothesis if my outcome is less than −6.1 or greater than +6.1. If the means of the samples differ by more than 6.1 minutes, I will conclude that the population means are probably not equal.

6 *Outcome*

My experiment outcome was $dm = m_1 - m_2 = 102 - 90 = 12$ minutes.

7 *Conclusion*

I reject the null hypothesis. We have good evidence that $\mu_1 > \mu_2$. The average dream time for Aquarians is longer than the average dream time for Capricorns.

SMALL-SAMPLE TESTS

When we compare two samples which are not both large, then the problem of how to describe the distribution of differences becomes complicated. To simplify the problem, several assumptions must be made about the two populations which are being compared. In this book we will treat only the situation which leads to the simplest calculations. In particular, we will assume that both populations are normally distributed to begin with.

In this case we can expect that the distribution of differences will be a t curve (in the same way that small samples led to a t curve before). We will assume that the particular t curve we want is the one with degrees of freedom equal to $n_1 + n_2 - 2$. (*Note:* There is discussion on a more advanced level which indicates that a more complicated formula is more accurate; but in many cases the two formulas give fairly similar results. So we will stick with $n_1 + n_2 - 2$ for simplicity.)

COMPUTER SIMULATION

We wrote a program to pick a random sample from each of the same two populations as before with $\mu_1 = 100$, $\mu_2 = 100$, $\sigma_1 = 10$, and $\sigma_2 = 8$. These two populations had been set up in the computer so that they were normally distributed. But this time we picked two *small* samples, where $n_1 = 20$ and $n_2 = 16$. The 100 results, arranged in numerical order, are shown in Table 8-4.

You will notice, if you compare this list of differences with the one at the beginning of this chapter where we took large samples, that these results are more scattered. In short, with small samples, the results are less predictable. In turn, this means that you cannot be confident about rejecting a hypothesis on the basis of small-sample evidence unless it is very strong evidence. That is why the critical t scores are larger than the critical z scores: results have to be further away from those predicted by the null hypothesis to be meaningful.

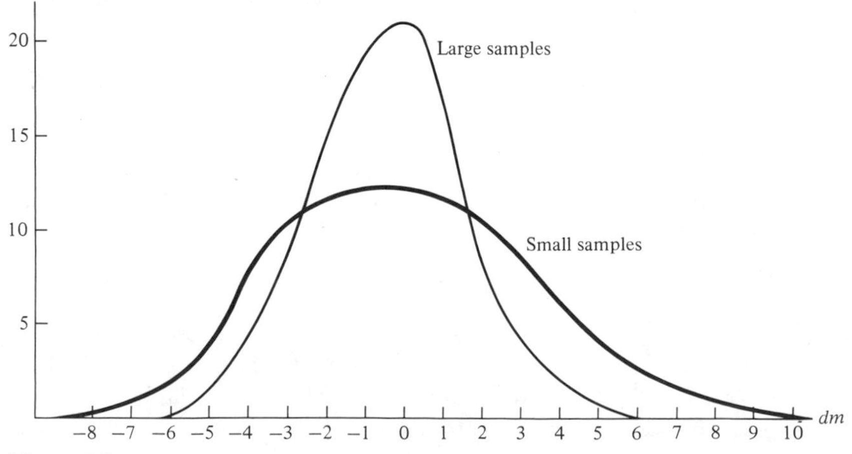

Figure 8-2

TABLE 8-4

Result Number	*dm*	Result Number	*dm*	Result Number	*dm*
1	−7.78	35	−1.46	68	1.69
2	−5.75	36	−1.45	69	1.70
3	−5.65	37	−1.25	70	1.77
4	−5.36	38	−1.11	71	1.82
5	−4.73	39	−0.90	72	1.86
6	−4.39	40	−0.93	73	1.88
7	−4.08	41	−0.80	74	2.09
8	−3.95	42	−0.85	75	2.12
9	−3.86	43	−0.78	76	2.18
10	−3.76	44	−0.67	77	2.27
11	−3.71	45	−0.63	78	2.47
12	−3.66	46	−0.53	79	2.47
13	−3.48	47	−0.28	80	2.54
14	−3.38	48	−0.10	81	2.91
15	−3.31	49	−0.19	82	2.95
16	−3.15	50	−0.09	83	3.11
17	−3.06	51	0.08	84	3.11
18	−2.96	52	0.24	85	3.48
19	−2.87	53	0.37	86	3.53
20	−2.62	54	0.39	87	3.70
21	−2.59	55	0.47	88	3.92
22	−2.56	56	0.51	89	4.08
23	−2.40	57	0.68	90	4.08
24	−2.30	58	0.72	91	4.10
25	−2.35	59	0.80	92	4.12
26	−2.24	60	0.85	93	4.19
27	−2.21	61	0.86	94	4.31
28	−2.09	62	1.01	95	5.07
29	−1.86	63	1.05	96	5.62
30	−1.82	64	1.29	97	6.89
31	−1.74	65	1.54	98	6.96
32	−1.73	66	1.54	99	8.35
33	−1.71	67	1.66	100	9.35
34	−1.63				

Figure 8-2 shows the smoothed graphs for the two sets of data. One is differences using large samples. The other is differences using small samples. You can see, for example, that in order to mark off the highest 5 percent on the small-sample curve, you have to draw a line much farther to the right than on the other graph.

CLASS EXERCISE

For the small-sample illustration summarized in Table 8-4:

8-10 Find the theoretical mean of the differences. $\mu_{dm} =$ ________. The actual mean of our 100 outcomes was 0.1.

8-11 Compute

$$\sigma_{dm} = \sqrt{\frac{\sigma_1^2}{n_1} + \frac{\sigma_2^2}{n_2}}$$

$\sigma_{dm} =$ __________. s_{dm} was calculated to be 3.2 in our simulation.

8-12 How many degrees of freedom are there? ____________________

8-13 If $\alpha = .05$, find t_c. ____________________

8-14 Compute $dm_c = 0 + (\pm t_c)(\sigma_{dm})$. $dm_c =$ __________ and __________.

8-15 Theoretically, what percentage of the *dm*'s do you expect to be in this interval?

____

8-16 What percentage of the actual *dm*'s are in this interval? ____________________

Example

Dr. U. Takimoff took a sample of 47 people entering Weight Losers Inc. After 4 weeks 25 of them were still attending meetings. The average of their 25 weights when they joined the program was 198 pounds, and the standard deviation was computed to be 20 pounds. Now their weights averaged 160 pounds with $s = 15$ pounds. Do a hypothesis test to see if the difference between these means is significant. Use $\alpha = .05$.

SOLUTION

1 *Populations*

$\text{pop}_1 =$ Weights of clients before coming to Weight Losers Inc.

$\text{pop}_2 =$ Weights after 4 weeks in the Weight Losers program.

2 *Hypothesis*

H_0 : There is no difference between the average weight of clients before and after participation.

$\mu_1 = \mu_2$ or $\mu_{dm} = 0$

$H_a : \mu_1 \neq \mu_2$ $\mu_{dm} \neq 0$

3 *Distribution*

Since $n_1 = n_2 = 25 \ngtr 30$, the distribution of differences is a t distribution. (Recall, we are assuming that weights are distributed normally which is a reasonable assumption.)

4 Sample data

$m_1 = 198$, $m_2 = 160$, $s_1 = 20$, $s_2 = 15$

$$s_{dm} = \sqrt{\frac{20^2}{25} + \frac{15^2}{25}} = \sqrt{\frac{400}{25} + \frac{225}{25}} = \sqrt{16 + 9} = \sqrt{25} = 5$$

Therefore, we have a td_{dm} ($\mu = 0$, $s = 5$).

5 Decision rule

df $= 25 + 25 - 2 = 48$. There is no entry in Table B-4 for 48 degrees of freedom. Therefore, we use the entry closest to 48. In our table the entry closest to 48 is 50. With $\alpha = .05$, we get $t_c = +2.01$.

$dm_c = 0 + (\pm 2.01)(5) = \pm\ 10.05$

Therefore, we will reject the null hypothesis if our difference is less than -10.05 pounds or greater than $+10.05$ pounds.

6 Outcome

Our experimental difference was $dm = 198 - 160 = 38$ pounds.

7 Conclusion

Since $38 > 10.05$, we reject the null hypothesis and conclude that the Weight Losers program really is effective.

CLASS EXERCISE

8-17 We claim that there is a difference in average grade between students who sit close to the teacher and those who sit far from the teacher.

a Split the class into two groups: the close ones and the far ones.

b Which group are you in? ____________________

c Write down the grade you got on the last test you took in this class.

____

d Find n, m, and s for each group.

Close group: $n_1 =$ ________ $m_1 =$ ________ $s_1 =$ ________

Far group: $n_2 =$ ________ $m_2 =$ ________ $s_2 =$ ________

$m_1 - m_2 =$ ____________________

e Using $\alpha = .05$, test to see if your class results support our claim.

CHAPTER SUMMARY

Differences between the means of pairs of samples from two populations are normally distributed if both samples are large. They have a t distribution if both are not large and are taken from normally distributed populations.

WORDS AND SYMBOLS YOU SHOULD KNOW

Difference of sample means
Distribution of differences of sample means

dm	s_{dm}
m_1	td_{dm}
m_2	μ_1
nd_{dm}	μ_2
s_1	μ_{dm}
s_2	σ_{dm}

ASSIGNMENT

A young statistician out west
While about a hypothesis test
Let α increase
Lest the experiment cease
For the rest of the test was the best

1 Two samples of humanoids are given a barfle exam. Six Martians scored 8, 5, 0, 4, 3, and 34. Five Venusians received scores of 20, 10, 50, 5, and −30.
  a Find n_1, m_1, and s_1.
  b Find n_2, m_2, and s_2.
  c Find s_{dm}.
  d Perform a hypothesis test to see if the results indicate a difference in the mean Martian barfle score and the mean Venusian barfle score. Use $\alpha = .01$. What are the two populations being compared?

2 The average score on a "helpfulness" evaluation for a random sample of 50 college administrators was 98 with $s = 10$. The average score for 50 maintenance personnel was 102 with $s = 20$. Test at the .05 significance level the hypothesis that there is no difference between the average score of administrators and the average score of maintenance personnel.

3 Joanne raises racing cockroaches. She is testing two diets to see if one is more effective than the other. Diet A is based on flour dissolved in Coca-Cola. Diet B is based on flour dissolved in Dr. Pepper. For the 10 roaches on diet A she finds that their average track speed is 13 inches per second with $s = 2$ inches per second. For the 10 roaches on diet B the average track speed is 15 inches per

second with $s = 3$ inches per second. Is this sufficient evidence at $\alpha = .01$ to be pretty sure that the diets are not equally effective?

4 Two types of gypsy moth traps were being tested. They were placed randomly in a small forest for 14 consecutive days. The type A traps caught an average of 194 moths per day with $s = 30$. Type B caught an average of 171 moths per day with $s = 27$. Is this sufficient at $\alpha = .01$ to indicate that one trap is more effective than the other?

5 In an experiment on rejection of foreign substances, skin grafts were made on mice. All the skin was taken from male mice. One batch of skin was grafted onto other male mice, another batch was grafted onto female mice. For the 20 male-male grafts, the mean survival time was 20 days with $s = 1.4$ days. For the 20 male-female grafts, the mean survival time was 14.5 days with $s = 1.3$ days. Formulate a hypothesis test based on this data. (Use $\alpha = .01$.) What are the *populations*?

6 In an experiment on causes of hunger, a group of rabbits was injected with chemicals. On some days they were given injections of a glucose solution; on others they got injections of a neutral saline solution. The time was measured from the beginning of the injection until the rabbit began to eat. For the 21 glucose injections, the average time was 9 minutes with $s = 1$ minute. For the 21 injections with neutral solution, the mean time before eating was 54 minutes with $s = 5$ minutes. Show that the probability that these two treatments are equally effective is less than .01.

7 Ecologists were conducting an experiment to see if the killing power of various bug killers changed when they were mixed with plant killers (as might happen in spraying crops). As part of the experiment, eight batches of 50 fruit flys were exposed to a solution of DDT, (a bug killer). On the average 8.5 died within 24 hours with $s = 7$. Then eight batches of 50 fruit flys were exposed to a mixture of DDT and Atrazine (a plant killer). On the average, 28.5 died within 24 hours with $s = 12$. Show that there is less than a 1 percent probability that this result happened just by chance, and therefore, show that probably the mixture of chemicals is more toxic than just DDT alone. (*Hint:* The value 50 in the problem just serves as a reference value. It does not enter the calculation any further.)

8 An experiment was being done to understand why the body of a pregnant woman does not reject a fetus since the attachment of the fetus is similar in many ways to an organ transplant. Various studies were done in the laboratory using appropriate tissue cultures. One protein (called HCG) which occurs naturally in pregnancy was noticed to have an effect which could block rejection. The blocking effect could be observed by noting a reduction in the amount of a chemical which develops to reject transplants. Five cultures were made with *no* protein HCG. The average amount of the rejecting chemical for these five cultures was 20 units with $s = 3$ units. Then five cultures were made with the amount of protein HCG corresponding to the amount present at pregnancy. The average amount of the rejecting chemical found was 15.6 units with $s = 2$ units. Is this a significant difference at $\alpha = .05$? What are the two populations being compared?

9 In an experiment on microscopic water animals, a scientist added vitamin E to the water to see what effect it might have on the growth of a particular species. She found 50 animals in water with vitamin E had a mean length of 630 units with $s = 8$ units. Then she found 80 animals in water without vitamin E had a mean length of 576 units with $s = 5$ units. Show that the probability is less than 1 percent that this difference was just due to luck.

10 At a committee meeting, an administrator claimed that students do not take the same number of credits in the fall semester as they do in the spring semester. A student member of the committee gathered a random sample of 10 students, asking each one how many credits he had carried last fall and how many he was carrying now. The figures were as follows:

Student	Fall	Spring
Alaric	15	12
Beowolf	18	18
Charlimain	12	13
Demosthenese	10	14
Ethelbert	14	12
Francowa	15	13
Giovanni	16	14
Hussar	14	12
Ivanovitch	12	12
Jeanne d'Arc	15	14

Do these figures confirm or deny the administrator's claim at the .05 significance level?

11 A teacher of shorthand is testing whether shorthand transcription speed varies between morning classes and afternoon classes. Using her class figures, perform a hypothesis test at the .05 significance level.

Rate of Speed	
A.M. Class	P.M. Class
120	120
130	130
120	120
110	110
100	110
110	110
110	90
120	110
	110
	120
	100
	120

12 If 100 radiator hoses made of rubber were found to have an average life of 1,000 hours with $s = 150$ hours and if 100 hoses made of plastic were found to have an

average life of 900 hours with $s = 200$ hours, answer these two questions.

a Which type is more consistent and, therefore, more predictable in its behavior?

b Is this evidence at the .01 significance level that the two types are not equal in average lifetime?

13 Two brands of automobile batteries are going to be compared. There will be 50 of each brand tested and the two sample means will be recorded. It is known that both brands have $\sigma = 200$ hours. How much difference does there have to be between the two sample means for us to decide they are not equal in average lifetime? Use $\alpha = .05$.

14 a Outline clearly a two-sample, sample-mean hypothesis test. Give the populations, samples, etc.

b Carry out the project that you have outlined in part a.

15 Make up your own question and use the data in Appendix A to answer it.

Readings from J. Tanur (ed.), "Statistics: A Guide to the Unknown," Holden-Day, San Francisco, 1972.

16 Drug Screening: The Never-Ending Search for New and Better Drugs, Dunnett, p. 23. Techniques for determining if a new drug is worthwhile for treating tumors in rats. Not all rats react the same way to the same drug. How can this be taken into account?

17 The Importance of Being Human, Howells, p. 92. What is the difference between a person and a chimp? The answer lies in the bones. What is a *discriminant function?*

18 Calibrating College Board Scores, Angoff, p. 237. How can the college board people say that a score of 534 in 1976 is equivalent to a score of 534 in 1973?

Review

19 The Legion of Mary, a traditional Catholic organization announced that a new experimental rosary with 20 decades would replace the centuries old 15-decade rosary. Virginia asked two groups of Catholics about their preference. The first group was asked if they preferred the new 20-decade rosary put out by the Legion of Mary. The second group was asked simply if they preferred the new 20-decade rosary. (No mention was made of the Legion of Mary.) Her data were as follows:

Group	Prefer New	Prefer Old	No Opinion
I	18	36	46
II	22	24	54

Perform a hypothesis test at the .05 significance level. What is the null hypothesis?

20 *U.S. World and News Report* stated that 36.7 percent of Americans smoke. One student's project was to determine if this figure applied to the people where he

was employed. A random sample of 100 employees was interviewed and 48 smoked. Test using $\alpha = .01$.

21 A student wanted to see if there was any difference between the probability that a female believes in the devil and the probability that a male does. The population was the employees and customers of Nick's Hardware Store. Out of 40 females interviewed, 17 believed in the devil, while out of 67 males, 35 believed. Using $\alpha = .01$, is this sufficient evidence to claim that more males believe in the devil than females do?

22 For his project, one student tested the claim in *Sports Illustrated* that the average man in America watches 5 hours of sports on a Sunday afternoon in the fall. The population that he tested was men who live in Massapequa, New York. The 10 men interviewed reported the following number of hours: 4.5, 3.0, 3.5, 4.0, 5.0, 3.5, 4.0, 4.0, 3.0, 4.5. Is this sufficient evidence to reject the claim that, on the average, men in Massapequa watch 5 hours of sports? Use $\alpha = .05$.

23 Test the claim that the average age of oak trees in a forest is 300 years if a sample of 60 trees has a mean of 326 years and s_{pop} of 125 years. Use $\alpha = .05$.

NINE

Estimation and Confidence Intervals

Every estimate ought to include an estimate of how much more it will cost than the estimate.

Anonymous

Hypothesis tests are like true-false questions:

30 percent of married women have jobs. $\pi = .30$. True or False?

The average height of Martians is 4.6 inches. $\mu = 4.6$ inches. True or False?

There is no difference between the proportion of males who smoke and the proportion of females who smoke. $\mu_{dp} = 0$. True or False?

There is no difference between the average life span of smokers and the average life span of nonsmokers. $\mu_{dm} = 0$. True or False?

Each of these null hypotheses states a claim. The researcher, on the basis of experimental data, either rejects the claim implying that it is probably false, or fails to reject it implying that it may be true. However, if we simply ask these questions, without supplying the numbers claimed (.30, 4.6 inches, 0, 0), they change from true or false declarative sentences to questions commonly known as "fill in the blanks." Thus, we ask questions such as:

What percentage of married women have jobs? $\pi =$ ________

What is the average height of Martians? $\mu =$ ________

What is the difference between the proportion of males who smoke and the proportion of females who smoke? $\pi_1 - \pi_2 =$ ___

What is the difference between the average life span of smokers and the average life span of nonsmokers? $\mu_1 - \mu_2 =$ ________

There is no claim. There is no hypothesis. We are not doing a statistical test of a claim. Instead we are *estimating* the value of a parameter.

POINT ESTIMATIONS

Ordinary common sense can lead you to figure out a way to estimate what percentage of married women have jobs. Simply gather a random sample of married women and ask. If 40 percent of your sample have jobs, then *about* 40 percent of the population also have jobs.

CLASS EXERCISE

Suppose you do gather a random sample of married women and find that 40 percent of them have jobs. Let $\pi = P$(a married woman in the population has a job).

9-1 On the basis of your sample data, can you conclude that $\pi = .40$? __________

Why? ____

9-2 $p = .40$. True or false? ____________________________

9-3 What is the difference in meaning between π and p? __________________

9-4 Gather a random sample of Martians and measure them. If the sample mean is 5 inches, then the mean of the population is near 5 inches.
How do we symbolize the two ideas in the last sentence? If __________ $= 5$ inches, then __________ is near to 5 inches.

Suppose we gather two random samples of adults, one of males and one of females. We compute the proportion of smokers in each sample. Suppose we find that $p_1 = .40$ while $p_2 = .35$. Then since $p_1 - p_2 = .05$, we estimate that $\pi_1 - \pi_2$ is *near* .05. Finally, we gather a random sample of the life spans of smokers and $m_1 = 71.3$ years. Then we gather a random sample of the life spans of nonsmokers. If $m_2 = 72.5$ years, then we conclude that $\mu_1 - \mu_2$ is *about* -1.2 years.

When we say that π is about 40 percent, we are giving a point estimate of π. A **point estimate** of a parameter is a single number which is our best guess for its value. In the example above we compute p, and then p serves as the point estimate for π.

When we use the value of m to estimate μ, then m is a point estimate for μ. In similar fashion we use the value of $p_1 - p_2$ to get a point estimate for $\pi_1 - \pi_2$. We use $m_1 - m_2$ as a point estimate of $\mu_1 - \mu_2$.

INTERVAL ESTIMATIONS

We now attempt to answer the questions: How near is near? Is 100 near 105?

If a student who paid $105 for a car tells a friend that it cost $100, that is not usually considered lying. We often tend to speak this way. We round off the exact value to a more convenient value which is close enough to still be meaningful. However, if a baby has 105° fever, we certainly would not tell the doctor it was near 100°!

In some situations if we estimate that π is about 33 percent, that might be very accurate, and the true value of π might be within 2 percent of our estimate, that is, $.31 < \pi < .35$. In another situation our estimate may be much less accurate, and the correct value of π might be as much as 10 percent away from p, that is, $.23 < \pi < .43$. Look at the number line below.

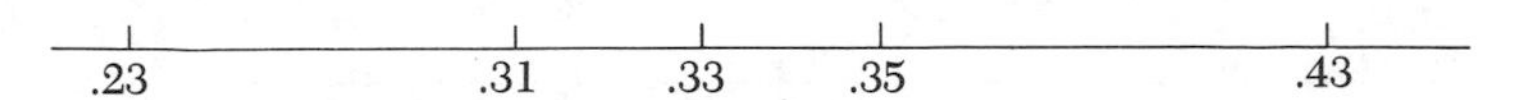

$p = .33$ corresponds to a point on the line, which is the reason that it is called a *point* estimate. We imply that the correct value of π is near 33 percent. .31 to .35 is an *interval* on the line. An estimate of the form $.31 < \pi < .35$ is called an **interval estimate.** We imply that we think π is more than .31 but less than .35.

CLASS EXERCISE

Have someone in the class set up two books any distance apart at the front of the class.

9-5 Guess the distance between the books. ______________________

9-6 Discuss your guess with others. Arrive at an interval estimate for the distance.

Between __________ and __________.

9-7 Are you 100 percent certain that answer 9-6 is correct? ______________

9-8 Are you *almost* positive that answer 9-6 is correct? ______________

The following is a quote from a newspaper printed in August 1973:

> Sickle cell anemia, a genetic disease, affects about 1 in every 500 blacks. Victims of the disease usually die prematurely. A far greater number of blacks—7 to 10 percent—have inherited the genetic trait from one parent but do not have the symptoms of the disease, because it develops only when the individual inherits the trait from both parents.

9-9 How do you suppose these two estimates were arrived at? ______________

9-10 Which one is a point estimate, "*1/500*" or "*from 7 to 10 percent*"?

____

9-11 Which one is an interval estimate? ______________________

CONFIDENCE INTERVALS

Let X equal the number of children in the families of students in this room. For you, X equals you plus your brothers and sisters. Let μ_X stand for the exact average number of children per family for all the students in the class. It is possible that $\mu_X = 3$.

I am 100 percent confident that $0 < \mu_X < 100$.
I am reasonably sure that $2 < \mu_X < 5$.

We wish to establish intervals in which we can be reasonably sure that a parameter lies. The smaller the interval is, the more useful it is.

If we are 95 percent sure that a parameter lies in an interval, we call that interval a **95 percent confidence interval.** If we have 99 percent assurance that the parameter is in the interval then we have a **99 percent confidence interval.**

Let us go back to our first example. What percent of married women have jobs? Let $\pi = P$(a married woman has a job). If in a random sample of 400 married women, 160 have jobs, then our point estimate for π is $p = 160/400 = .40$.

Back in Chapter 5 when discussing one-sample binomial hypothesis tests, our null hypothesis assumed that we knew the value of π. Now we have no such assumption. Since we do not know the value of π, we cannot check to see if $n\pi$ and $n(1 - \pi)$ are both larger than 5. Neither can we compute

$$\sigma_p = \sqrt{\frac{\pi(1-\pi)}{n}}$$

However, p is an estimate for π. Therefore, we can estimate these numbers as follows:

Parameter	Estimate
$n\pi$	np
$n(1-\pi)$	$n(1-p)$ or nq
$\sigma_p = \sqrt{\frac{\pi(1-\pi)}{n}}$	$s_p = \sqrt{\frac{p(1-p)}{n}} = \sqrt{\frac{pq}{n}}$

Similarly, in a two-sample binomial hypothesis test our null hypothesis assumed that $\pi_1 = \pi_2$, and we called this common value π. Now we have no such assumption of equality. Hence, instead of using

$$s_{dp} = \sqrt{\frac{pq}{n_1} + \frac{pq}{n_2}}$$

we shall use

$$s_{dp} = \sqrt{\frac{p_1q_1}{n_1} + \frac{p_2q_2}{n_2}}$$

Illustration

The basic problem of confidence intervals is this: We have a sample statistic. We ask how far away might this statistic be from the true parameter? We can answer this question if we know what values of the statistic are common in repeated experiments of this type. Recall that in our random sample of 400 married women, 40 percent had jobs. So $p = .40$. We ask, "If p is .40, how far away can this be from the *true* percent of *all* married women who have jobs?" We have $np = 400(.40) = 160$ and $nq = 400(.60) = 240$. Since 160 and 240 are both bigger than 5, the distribution of p's is approximately normal. Note that 160 and 240 are the outcomes of our sampling: 160 did have jobs and 240 did not. It is always the case in a one-sample binomial confidence interval that np and nq equal the outcomes.

$$s_p = \sqrt{\frac{pq}{n}} = \sqrt{\frac{(.4)(.6)}{400}} = \sqrt{\frac{.24}{400}} = \sqrt{.0006} = .0245$$

Therefore, in repeated sampling of married women we have $nd_p(\mu = \pi,\ s = .0245)$, and our value of $p = .40$ is one member of this distribution. (See Figure 9-1.)

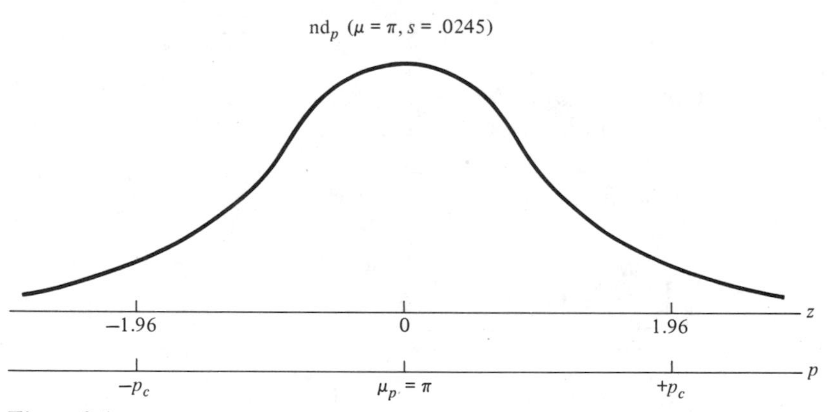

Figure 9-1

Since 95 percent of all the p's are between the two critical p's, there is a .95 probability that *our* p, .40, is between the two critical values, p_c. Recall that $p_c = \mu_p + zs$. Therefore,

$$p_c = \pi + (\pm 1.96)(.0245)$$
$$p_c = \pi \pm .048$$

(See Figure 9-2.)

We can be 95 percent certain that the difference between our $p = .40$ and the true value of π is less than .048, which means we can be 95 percent certain that the true value of π is between .352 and .448. We call the interval from .352 to .448 the 95 percent confidence interval for π, and write $.352 < \pi < .448$.

Note that this came from $.40 - .048 < \pi < .40 + .048$, where

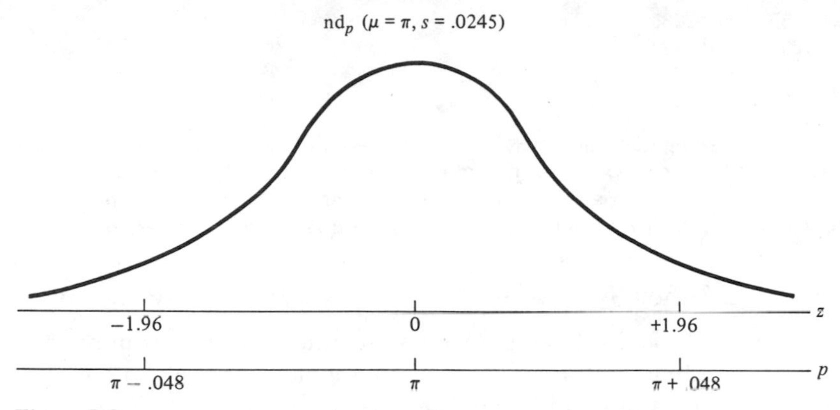

Figure 9-2

$.40 = p$ and $.048 = zs$. Thus, the formula for this confidence interval is

$$p - zs < \pi < p + zs$$

where the value of z depends on the degree of confidence you want. For 95 percent confidence, use $z = \pm 1.96$ and for 99 percent confidence, use $z = \pm 2.58$.

COMPUTER SIMULATION

Back in Chapter 5, we had a computer pick many random samples of 200 students from a population in which we had set $\pi = .30$, where π was the true percentage of students handing in phoney book reports.

The first five values of p were:

.315
.32
.36
.29
.285

If we compute a 95 percent confidence interval using the first $p = .315$, we have

$$s = \sqrt{\frac{(.315)(.685)}{200}} = \sqrt{.00107875} = \sqrt{.00108} = .03286$$

and

$$zs = (1.96)(.03286) = .064$$

Thus our 95 percent confidence interval is

$$p - zs < \pi < p + zs$$
$$.315 - .064 < \pi < .315 + .064$$
$$.251 < \pi < .379$$

We would say, on the basis of this experiment, that we are 95 percent sure that π is between .251 and .379.

The other four values of p lead to the following four 95 percent confidence intervals:

$.255 < \pi < .385$
$.293 < \pi < .427$
$.227 < \pi < .353$
$.222 < \pi < .348$

Notice that the true value of π, .30, is within each of these five intervals. This will happen 95 percent of the time. Or to put it another way, the probability is .95 that we have trapped the true value of π in such an interval. Similarly, when you compute a 99 percent confidence interval, the probability is .99 that you will trap the true value of the parameter.

Other Situations

Just as the above one-sample binomial experiment leads to a confidence interval, so in an analogous way we can construct different types of confidence intervals in different situations.

Notice that our formula

$$p - zs < \pi < p + zs$$

says that to estimate a parameter, first get the point estimate and then add to and subtract from that number an error term. This has the basic format

$$\text{Estimate} - \text{error} < \text{parameter} < \text{estimate} + \text{error}$$

We can derive similar formulas for sample means, both large and small samples. We can derive formulas for both one- and two-sample situations. These formulas are summarized in Table 9-1 and illustrated in the following examples.

Examples of Confidence Intervals

1 Find the 99 percent confidence interval for the difference between the percentage of freshmen and the percentage of sophomores who know that "impeachment" does not mean "removal from office," if 40 freshmen out of 100 surveyed and 100 sophomores out of 200 surveyed knew the difference.

Solution

Let population 1 be the freshmen and population 2 be the sophomores.

$$p_1 = 40/100 = .40 \qquad p_2 = 100/200 = .50$$

Therefore, the point estimate for the difference is $p_1 - p_2 = .40 - .50 = -.10$.

$$n_1p_1 = \frac{100(40)}{100} = 40 \qquad n_1q_1 = \frac{100(60)}{100} = 60$$

$$n_2p_2 = \frac{200(100)}{200} = 100 \qquad n_2q_2 = \frac{200(100)}{200} = 100$$

TABLE 9-1 SUMMARY OF ESTIMATING PROCEDURES

	Parameter	Point Estimate	Conditions	Confidence Interval	Formulas
I Binomial One-Sample	π	p	Two outcomes greater than 5	$p - zs_p < \pi < p + zs_p$	$s_p = \sqrt{\frac{pq}{n}}$ °
II Binomial Two-Sample	$\pi_1 - \pi_2$	$p_1 - p_2$	Four outcomes greater than 5	$(p_1 - p_2) - zs_{dp} < \pi_1 - \pi_2 < (p_1 - p_2) + zs_{dp}$	$s_{dp} = \sqrt{\frac{p_1q_1}{n_1} + \frac{p_2q_2}{n_2}}$ °
III A Sample-Mean One-Sample, Large Sample	μ	m	$n > 30$	$m - zs_m < \mu_{pop} < m + zs_m$	$s_m = \frac{s_{pop}}{\sqrt{n}}$
III B Sample-Mean One-Sample, Small Sample	μ	m	$n \ngtr 30$ population is normal	$m - ts_m < \mu_{pop} < m + ts_m$	$\text{df} = n - 1$
IV A Sample-Mean Two-Sample, Large Sample	$\mu_1 - \mu_2$	$m_1 - m_2$	$n_1 > 30$ and $n_2 > 30$	$(m_1 - m_2) - zs_{dm} < \mu_1 - \mu_2 < (m_1 - m_2) + zs_{dm}$	$s_{sm} = \sqrt{\frac{s_1^2}{n_1} + \frac{s_2^2}{n_2}}$
IV B Sample-Mean Two-Sample, Small Sample	$\mu_1 - \mu_2$	$m_1 - m_2$	$n_1 \ngtr 30$ or $n_2 \ngtr 30$ both populations are normal	$(m_1 - m_2) - ts_{dm} < \mu_1 - \mu_2 < (m_1 - m_2) + ts_{dm}$	$\text{df} = n_1 + n_2 - 2$

° These two formulas for binomial confidence intervals are different from the ones used for hypothesis testing.

These four numbers are exactly the same as the sample results: 40 freshmen knew, 60 freshmen did not know; 100 sophomores knew, and 100 sophomores did not know. This is always the case in a two-sample binomial confidence interval.

Since 40, 60, 100, 100 are all greater than 5, we can assume that we have a normal distribution of differences, *dp*, with mean equal to $\pi_1 - \pi_2$ and

$$s_{dp} = \sqrt{\frac{p_1q_1}{n_1} + \frac{p_2q_2}{n_2}}$$
$$= \sqrt{\frac{(.4)(.6)}{100} + \frac{(.5)(.5)}{200}} = \sqrt{.0024 + .00125} = \sqrt{.00365} = .0604$$

Thus we have

$$(p_1 - p_2) - zs < \pi_1 - \pi_2 < (p_1 - p_2) + zs$$
$$-.10 - 2.58(.0604) < \pi_1 - \pi_2 < -.10 + 2.58(.0604)$$
$$-.10 - .16 < \pi_1 - \pi_2 < -.10 + .16$$
$$-.26 < \pi_1 - \pi_2 < +.06$$

This means that the true difference could be anywhere from 26 percent more sophomores than freshmen knowing to 6 percent more freshmen than sophomores knowing that impeachment does not mean removal from office.

2 Find the 95 percent confidence interval of the average weight of statistics teachers if a random sample of 25 statistics teachers had a mean weight equal to 145 pounds and a standard deviation (s_{pop}) equal to 15 pounds. It is reasonable to assume that the population of weights is normally distributed.

Solution

Since $n = 25 \ngtr 30$, we assume a t distribution of the means of samples of 25 teachers.

$$\text{df} = 24$$

$$t_c = \pm 2.06$$

$$s_m = \frac{s}{\sqrt{n}} = \frac{15}{\sqrt{25}} = 3$$

$$m - t_c s_m < \mu_{\text{pop}} < m + t_c s_m$$
$$145 - 2.06(3) < \mu < 145 + (2.06)(3)$$
$$145 - 6.2 < \mu < 145 + 6.2$$
$$138.8 < \mu < 151.2$$

Thus we are 95 percent certain that the average weight of all statistics teachers is between 138.8 and 151.2.

3 Find the 95 percent confidence interval for the difference between the average weight of Jupiter's mosquitos and the average weight of Saturn's mosquitos if a sample of 100 mosquitos from Jupiter had an average weight of 48 ounces with a standard deviation of 10 ounces, while a sample of 100 mosquitos from Saturn had an average weight of 35 ounces with a standard deviation of 12 ounces.

Solution

$n_1 = 100$, $n_2 = 100$. Both are greater than 30, so we can assume a normal distribution of differences, dm.

$$s_{dm} = \sqrt{\frac{s_1^2}{n_1} + \frac{s_2^2}{n_2}} = \sqrt{\frac{100}{100} + \frac{144}{100}} = \sqrt{1 + 1.44} = \sqrt{2.44} = 1.56$$

$$(m_1 - m_2) - zs_{dm} < \mu_1 - \mu_2 < (m_1 - m_2) + zs_{dm}$$
$$13 - 1.96(1.56) < \mu_1 - \mu_2 < 13 + 1.96(1.56)$$
$$13 - 3.06 < \mu_1 - \mu_2 < 13 + 3.06$$
$$9.94 < \mu_1 - \mu_2 < 16.06$$

We are 95 percent certain that the mosquitos from Jupiter are between 9.94 ounces and 16.06 ounces heavier than those from Saturn.

Chapter Summary

We have distinguished between hypothesis tests and problems of estimation. A point estimate is that single value, based on sample data, which is used to estimate some population parameter. An interval estimate is a range of values, based on sample data, which we expect to include the true value of some parameter. When we attach a probability measure to an interval estimate, we have a confidence interval.

Words and Symbols You Should Know

Confidence interval
Interval estimate
Point estimate

ASSIGNMENT

A confident professor just "knew"
How to estimate square root of two
It's somewhere between
Point 1 and sixteen
But he couldn't get closer, can you?

1 Identify the following statistical situations as one of the following:

A Binomial one-sample hypothesis test
B Binomial one-sample estimation
C Binomial two-sample hypothesis test
D Binomial two-sample estimation
E Sample-mean, one-sample hypothesis test
F Sample-mean, one-sample estimation
G Sample-mean, two-sample hypothesis test
H Sample-mean, two-sample estimation
I Other

a What percentage of college women smoke?
b Do 30 percent of doctors smoke?
c There is no difference between the IQs of rich people and poor people.
d The average height of midgets is 2 feet 11 inches.
e What is the difference between the cumulative grade point averages of veterans and nonveterans?
f If you know the weight of a 7-year-old boy, can you estimate his height?
g Is the percentage of black high school graduates who earn college degrees the same as the percentage of whites?
h What is the difference between the average number of fleas on a red squirrel and the average number of fleas on a gray squirrel?
i Find the difference between the percentage of Martians who do and the percentage of Venusians who do.

2 a Find five 99 percent confidence intervals for $\pi_1 - \pi_2$ based on the data from the computer simulation in Chapter 6. The data is repeated here:

$$n_1 = 40 \qquad n_2 = 50$$

Experiment Number	p_1	p_2	$dp = p_1 - p_2$
1	.725	.780	−.055
2	.730	.820	−.120
3	.800	.740	+.060
4	.825	.800	+.025
5	.775	.800	−.025

b Recall that $\pi_1 - \pi_2 = .80 - .80 = 0$. Did we capture this parameter in each confidence interval?

3 a Find the 95 percent confidence interval for μ based on the data from the first large sample computer simulation in Chapter 7. For that data we have

$$n = 36 \qquad m = 151.8 \text{ pounds}$$

and s is computed to be 13.2.

b Find the 95 percent confidence interval for μ based on the second large-sample computer simulation in Chapter 7. The data is repeated here.

133.4811	143.6725	156.0080	152.5119	158.0724	157.5451
161.7672	155.5889	142.4322	145.5940		
167.8342	129.2312	158.5183	149.8359	147.2219	154.8697
159.4607	153.4592	152.7085	156.2788		
137.4518	138.0017	136.3693	135.3579	144.6839	166.4477
150.5138	154.0063	134.2868	161.7211		
148.1546	155.5299	138.2188	125.6142	141.6015	142.7654

c Find the 95 percent confidence interval for μ based on the data from the first small-sample computer simulation in Chapter 7. For that data we have

$$n = 16 \qquad m = 149.9$$

and s can be computed to be 17.7.

d Find the 95 percent confidence interval for μ based on the data from the second small-sample computer simulation in Chapter 7. The data is repeated here.

135.726	157.623	163.447	162.252	176.596	169.499	147.745
158.193	152.975	174.651	148.457	138.388	166.694	149.557
151.845	127.12					

e Did the four confidence intervals above contain the population parameter $\mu_{pop} = 150$ pounds?

4 Find the 99 percent confidence interval for $\mu_1 - \mu_2$ based on the data from the computer simulations in Chapter 8.

a $n_1 = 50$, $m_1 = 102.01$, $s_1 = 10.2$, $n_2 = 32$, $m_2 = 100.15$, $s_2 = 7.9$

b $n_1 = 20$, $m_1 = 96.54$, $s_1 = 9.9$, $n_2 = 16$, $m_2 = 104.32$, $s_2 = 8.3$

c Is the parameter $\mu_1 - \mu_2 = 0$ within both of these confidence intervals?

5 Timothy O'Flaherty plays Boccie regularly. In the past year he has won 80 games out of the 100 he played. Compute the 95 percent confidence interval for the percentage of games he should win next year based on the data from this year. Are we 95% sure that he will win more than 75 percent of his games next year?

6 a Estimate the percentage of blacks who carry the sickle cell trait in their blood if a random sample of 579 blacks showed 39 with the sickle cell trait.

b Find a 99 percent confidence interval for your estimate.

c Compare these results, with the newspaper report quoted immediately after Exercise 9-8.

7 A random sample of 345 adults in Stochaic county in connection with the new drug laws showed:

45—no opinion
130—disagree with new laws
170—favored new law

a Estimate the percentage of people in favor of the new drug law.

b Find a 99 percent confidence interval for your estimate.

8 In order to estimate the percentage of customers at Clarks who reside in Clarksville, 100 customers were interviewed. It was found that 63 lived in Clarksville.
   a Estimate the percentage of all customers who reside in Clarksville.
   b Find the 95 percent confidence interval for your estimate.
9 Of 50 women interviewed, 43 favored state-operated Day Care Centers. Of 50 men interviewed, 44 favored the centers.
   a Estimate the difference between the percentage of men who favor state-operated Day Care Centers and the percentage of women who do.
   b Find the 99 percent confidence interval for your estimate.
10 A student asked 180 persons their opinion about women wearing pants to work. Out of 90 she asked while wearing pants, 64 approved. Of the 90 she asked while wearing a dress, only 40 approved.
   a Estimate the difference between people reacting to this question when the interviewer wore pants and when the interviewer wore a dress.
   b Find the 99 percent confidence interval for your estimate.
11 To estimate the television-watching habits of customers for advertising purposes, a survey of customers at M. H. Racy's was taken. Of 72 males interviewed, 52 watched television for more than 5 hours a week. Out of 81 women interviewed, 60 watched television for more than 5 hours a week. Find the 95 percent confidence interval for the difference between the percentage of males and the percentage of females who watched television for more than 5 hours a week.
12 A Gallup poll taken in August 1973 claimed that 48 percent of Americans opposed premarital sex. Random samples in Martin's Ferry produced the following figures: 23 out of 40 males and 27 out of 40 females opposed premarital sex. Estimate the difference between the opinions of the sexes and find a 95 percent confidence interval for your estimate.
13 a Estimate the average height of 17-year-old girls in the United States if a random sample of 40 girls had an average height of 63.8 inches with $s_{\text{pop}} = 2.2$ inches.
   b Find a 95 percent confidence interval for your estimate.
14 a Estimate the average age of men joining local volunteer fire departments if a random sample of 20 new recruits had a mean age of 21.2 years with a standard deviation of 2.6 years.
   b Find the 95 percent confidence interval for your estimate.
15 Find the 99 percent confidence interval for the average salary of college librarians if a sample of 49 librarians tested at a recent national convention produced $m = \$13{,}000$, $s = \$1{,}400$.
16 a Would the interval in Assignment 15 become wider or narrower if $s = \$7{,}000$?
   b Would the interval in Assignment 15 become wider or narrower if $s = \$1{,}400$ but $n = 16$?
   c Compute these two intervals. Were your predictions in parts a and b correct?
17 If 23 soldiers averaged 4.3 wounds with $s_{\text{pop}} = 6$, find the 99 percent confidence interval for the average number of wounds.

18 Tomorrow 500 boys are going on a picnic. A random sample of 36 boys shows that they will consume the following number of cans of soda each: 3, 5, 0, 2, 4, 3, 2, 12, 2, 3, 5, 4, 0, 6, 4, 3, 3, 4, 2, 0, 1, 1, 5, 4, 3, 2, 5, 14, 2, 3, 3, 2, 2, 4, 1, 3.

a Find the 95 percent confidence interval for the average number of cans.

b How many cans of soda should we buy if we want to be 95 percent certain that all 500 boys will have enough?

19 Compute the 99 percent confidence interval for the average number of television sets per household in the village of Volley Stream if a random sample of homes yielded the following numbers of television sets per household: 4, 2, 1, 4, 3, 2, 4, 5, 3, 3, 3, 4, 0, 1, 1, 2, 2, 1, 4, 5, 3, 2, 0, 1, 2.

20 Is there any difference between the average age of Colonial Maid Ice Cream customers and what the Colonial Maid truck drivers think the average age is? A random sample of 100 customers had a sample mean equal to 12.3 years with a standard deviation of 8.2 years. A sample of 32 truck drivers' estimates had a mean of 9.8 years with $s = 2.6$.

a Find the 95 percent confidence interval for this difference.

b Comment on this remark by one experimenter: "Drivers tend to think of the modal age rather than the mean age."

21 Estimate the difference between the average number of children a male student at State U intends to have and the average number of children a female student intends to have. A random sample of 31 males had an average 2.6 with $s = 1.8$. A sample of 33 females averaged 2.3 with $s = 1.6$. Find the 99 percent confidence interval for your estimate.

22 A group of 20 adults on Dr. Quietman's diet lost, in 1 month, an average of 10.2 pounds with $s = 2.2$ pounds. A sample of 12 adults on Dr. Oddkin's diet averaged, in 1 month, an 8.1-pound loss with $s = 1.9$. Find the 99 percent confidence interval for the difference in weight loss.

23 A random sample of the ages of elderly people in retirement homes showed that for 40 males $m = 80.1$, $s = 6.3$, while for 20 females $m = 81.2$ and $s = 7.0$. Find the 95 percent confidence interval for the difference.

24 To test the psychological effects of aspirin, two groups of people were given "aspirin" for headaches. Some really were given aspirin, while others were given a similar looking pill containing no aspirin. The length of time it took for the headaches to go away was timed. In the group given aspirin, 500 persons averaged 4.8 minutes with $s = 2$ minutes. In the other group, 500 persons averaged 6.0 minutes with $s = 3$ minutes. If μ_1 equals the average length of time for headaches to disappear with aspirin and μ_2 equals the average length of time for headaches to disappear without aspirin, find the 99 percent confidence interval for $\mu_1 - \mu_2$.

25 In a hospital, 20 boys had an average of 9.7 fingers with a standard deviation of .2, while 20 girls had an average of 9.9 fingers with a standard deviation of .1. Find the 95 percent confidence interval for this difference.

26 A random sample of 20-year-olds who have been driving for at least 3 years yielded the following data concerning the total number of accidents in the last 3 years:

BOYS	GIRLS
3, 0, 1, 1, 0	2, 1, 3, 3, 1
0, 0, 2, 2, 2	3, 0, 0, 0, 0
1, 2, 4, 2, 1	2, 7, 1, 0, 2
3, 0, 6, 2, 1	5, 3, 0, 0, 1
0, 1, 3, 2, 3	0, 1, 2, 3, 0
1, 3, 0, 1, 0	0, 1, 1, 2, 1
2, 3, 2, 1, 2	0, 1, 0, 1, 1

a Find the 95 percent confidence interval for the difference between the average number of accidents.

b Are you 95 percent sure that boys average more accidents than girls do?

27 Are history majors more aware of current events than others? A test of current events was given to two groups: 50 history majors had an average test score of 73 with a standard deviation of 10; 20 non-history majors averaged 63 with $s = 10$. Find the 99 percent confidence interval for the difference between these averages.

28 An article in a science magazine states that $\pi = P$(a boy is born) is 51.6 percent. The author states that this figure is based on a random sample of 4,000 births. Could it be 50 percent?

a Find the 95 percent confidence interval.

b Estimate the probability that π could be as low as 50 percent.

c Find the 95 percent confidence interval if n were only 1,000.

29 Refer to the thumbtack problem in Assignment 6 of Chapter 3 and the note to the instructor. Graph the data in Figure A9-29. Find the 95 percent confidence interval for $\pi = P$(point lands up).

30 Dr. Cy N. Test noted that mutant nematodes whose heads were not aligned with their bodies tended to move in spirals toward a chemical attractant, showing that nerves in the head of the worms detected the chemical. (Normal worms move on straighter paths.) Of 50 such mutants tested, 44 traveled in a spiral path.

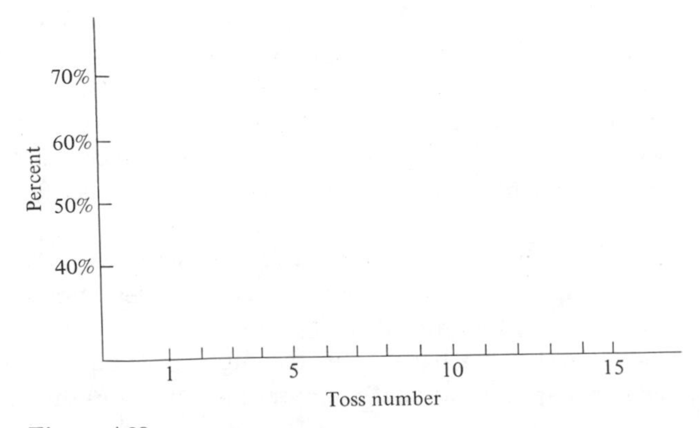

Figure A29

Estimate the percentage of such mutants that travel in a spiral, and find a 95 percent confidence interval for this percentage.

31 On August 11, 1974 *The New York Times* (page 44) stated that 79 percent of a random sample of "550 American voters" backed the President's resignation. The paper concluded that "more than three-quarters of Americans" backed the President's resignation.

a Compare the article in *Newsweek*, August 19th, which refers to a *sample* of 550 American voters and the article in *The New York Times* which refers to the *population* of Americans.

b Find the 95 percent confidence interval for the estimate.

c Find the 99 percent confidence interval for this estimate.

d Comment on the conclusion in *The New York Times* concerning "Americans" based on this "sample of Americans."

32 a Outline an experiment that you could perform to estimate a population parameter and to find a confidence interval for your estimate. Give details clearly.

b Perform the experiment that you have described in part a.

33 Make up your own question and use the data in Appendix A to answer it.

34 Read about estimates and intervals in Chapter 8 of *How to Take a Chance* by Darell Huff.

Readings from J. Tanur (ed.), "Statistics: A Guide to the Unknown," Holden-Day, San Francisco, 1972.

35 Election Night on Television, Link, p. 137. Estimating final results from early returns.

36 Looking through Rocks, Chayes, p. 367. How to estimate the percentage of a rocky region made up of a particular mineral.

37 How to Count Better: Using Statistics to Improve the Census, Hansen, p. 276. How good is an estimate based on only 20 percent of the population? What is an enumerator?

Review

38 Does knowing someone who has had an abortion affect one's attitude about abortion laws? A random sample of people was questioned and the following responses were obtained:

Person Questioned	Favor Stricter Laws	Favor Present Laws	Favor Easier Laws	No Opinion
Knows Someone Who Had An Abortion	15	25	10	0
Does Not Know Someone Who Had An Abortion	33	59	34	24

Test using $\alpha = .05$. State in words what the null hypothesis is.

39 A statistics student employed part-time in an automotive parts store doubted the

advertised claim that 60 percent of private individuals who bought GM cars, which come with brand *A* spark plugs, switched over to brand *B* spark plugs. He sampled 200 male customers who owned cars that came with brand *A* spark plugs and who do their own tune-ups; 109 said that they used brand *B*. Is this sufficient evidence at the .05 significance level to reject the advertisement?

40 In the midst of opposition to a new shopping center, a student interviewed 100 local residents under 31 years old and 100 who were 31 or older. Of these, 43 of the younger group and 62 of the older group opposed the shopping center. Is this sufficient evidence at the .01 significance level to show that the two age groups differ on this issue?

41 A management official at Gron's tells employees that the average part-time night employee earns $2.30 per hour and that forming a union would cost them more than it could gain for them. Nan sampled 12 other fellow employees and got the following figures:

1 Nan	$2.10	8 Jean	$2.15
2 Joanne	2.10	9 Marian	2.05
3 Eileen	2.00	10 Debbie	2.20
4 Grace	2.00	11 Kathy	2.75
5 Ann	2.00	12 Betty	2.40
6 Mark	2.00	13 Cathy	2.50
7 Tom	2.00		

Does this data tend to disprove the management's statement at the .05 significance level?

42 A survey to find out the average number of hours that students worked each week produced the following data: For 50 male students, $m_1 = 17.1$ hours and $s_1 = 15.1$ hours. For 50 female students $m_2 = 8.6$ hours and $s_2 = 10.5$ hours. Test at the .05 significance level to see if there is a difference between the means of the populations.

TEN

Correlation and Prediction

A dog starved at the gate predicts the
ruin of the state.

William Blake

CLASS EXERCISE

10-1 My son is 3 years old. Estimate his weight. ______

10-2 Predict his height at age 12. ______

Suppose that Alice reads more books than Barbara does. Would you therefore expect any of the following to be true? (Would you like to make a monetary wager on any of them?)

10-3 Alice is taller than Barbara. ______

10-4 Alice is taking more college credits than Barbara. ______

10-5 Alice gets better grades in history than Barbara. ______

10-6 Alice has a higher grade point average than Barbara. ______

10-7 Alice has more brothers than Barbara. ______

Suppose Adam is heavier than Bruce. Which of the following might you expect to be true?

10-8 Adam is taller than Bruce. ______

10-9 Adam eats more than Bruce. ______

10-10 Adam is older than Bruce. ______

10-11 Would your answers change if you knew the boys were brothers in grade school? ______

The question that we are asking is, when are two variables related? Is there a correlation between the age and the height of children?

10-12 If you arranged young children at a family reunion by age, would they also be in order by height? ______

10-13 Perfectly? ______

10-14 Near perfect? ______

SCATTERGRAM

In one group of people we might have age and height as indicated by the data in Table 10-1. In another group we have the data shown in Table 10-2.

If we draw two graphs called **scattergrams** for these data, we get the results shown in Figures 10-1 and 10-2. Each dot on the scattergram represents one person. For instance, Cathy was 6 years old and 46 inches tall, so a dot was placed in the spot that is

TABLE 10-1

Person	Age	Height, Inches
Cathy	6	46
Donna	7	47
Ed	8	50
Frank	8	51
Gerri	10	53

TABLE 10-2

Person	Age	Height, Inches
Henry	16	73
Ivan	16	71
Janet	17	60
Karen	17	66
Leroy	18	64

directly over 6 and exactly to the right of 46 inches. Notice the patterns. In the first graph the dots are somewhat in a straight line. The dots in the second graph do not form such a linear pattern. When the dots do form a straight-line pattern, we say that there is **linear correlation** between the two variables we are discussing. When this type of correlation exists, it can be used to make predictions about one variable based on what we know about the other one.

The British statistician, Karl Pearson, determined a way to measure how close a series of points comes to forming a perfect straight line. This measure is called the **Pearson coefficient of correlation** or simply the **correlation coefficient.** If the points on the scattergram actually lie on a straight line, the correlation coef-

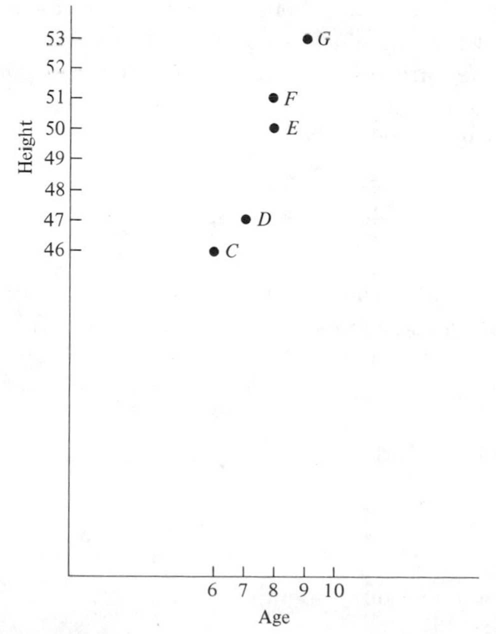

Figure 10-1

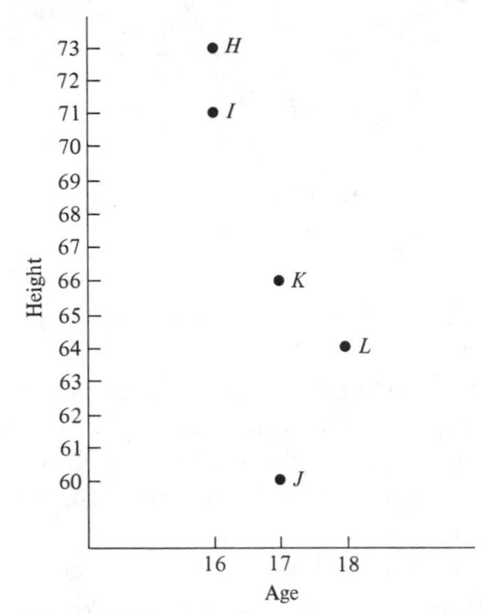

Figure 10-2

ficient equals +1 or −1. The coefficient of correlation is positive if the line goes up to the right and negative if the line goes down to the right. If the pattern of points is close to being a straight line, the correlation coefficient might be ±.9 or ±.8. As the pattern points become less and less linear, the correlation coefficient gets nearer and nearer to zero. When the pattern of dots is completely random and the correlation coefficient equals zero, we say that the two variables are *uncorrelated.* When two variables are uncorrelated knowledge about one of them is of no use in making predictions about the other.

CLASS EXERCISE

Guess the value of the coefficient of correlation between the following variables. (Possible answers: +1, high positive, some positive, zero, some negative, high negative, or −1.)

10-15 The variables graphed in Figure 10-1. ______

10-16 The variables graphed in Figure 10-2. ______

10-17 Height and weight of children. ______

10-18 Height and weight of adults. ______

10-19 Annual income and amount of money spent on food. ______

10-20 Annual income and percent of income spend on food. ______

10-21 Years of schooling and years spent in prison. ____________________

10-22 Height and number of pets owned. ____________________

HOW TO COMPUTE THE COEFFICIENT OF CORRELATION

Suppose we want to see if there is any correlation between scores on a math test and scores on a physics test, as shown by the following data.

Student	Math Test	Physics Test
Marvin	9	20
Norma	8	21
Otto	8	18
Paddy	7	16
Quentin	5	16
Raoul	5	14
Samantha	4	17

If these seven people are the entire population under consideration, then we can compute the correlation coefficient, and we symbolize it by the Greek letter ρ (rho). More usually, however, we would not have information for an entire population but only for a random sample taken from it. In this case we estimate the value of ρ and symbolize it by the English letter r.

There are several methods of computing the value of r. One easy way is with the formula

$$r = \frac{n\Sigma XY - (\Sigma X)(\Sigma Y)}{\sqrt{n\Sigma X^2 - (\Sigma X)^2}\ \sqrt{n\Sigma Y^2 - (\Sigma Y)^2}}$$

where X stands for the values of the first variable, Y represents the value of the other variable, and n is the number of pairs of values in the sample. Example: We are going to compute the coefficient of correlation for the data above considering it to be a random sample from a larger population.

CLASS EXERCISE

10-23 Before you start guess the answer. $r =$ ____________________
(Recall that r is a number between -1 and $+1$ inclusive.)

10-24 You could improve your guess by sketching a scattergram. So, after you have answered Exercise 10-23, sketch the scattergram in Figure CE10-24.

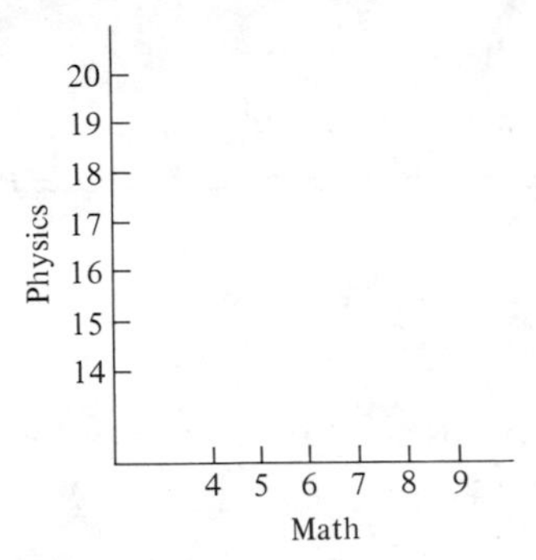

Figure CE10-24

10-25 After looking at your graph, revise, if necessary, your guess for the value of the coefficient of correlation. $r =$ ____________________

Now let us compute r. Recall the data,

X	Y
9	20
8	21
8	18
7	16
5	16
5	14
4	17

and the formula

$$r = \frac{n\Sigma XY - (\Sigma X)(\Sigma Y)}{\sqrt{n\Sigma X^2 - (\Sigma X)^2}\ \sqrt{n\Sigma Y^2 - (\Sigma Y)^2}}$$

To use this formula, we need

ΣXY
ΣX
ΣY
ΣX^2
ΣY^2
n

Only the first of these symbols, ΣXY, represents a new idea. It says, multiply X and Y and add them up. Essentially we need the following numbers.

X	Y	X^2	Y^2	XY
9	20			
8	21			
8	18			
7	16			
5	16			
5	14			
4	17			

Filling in the appropriate columns and adding, we get

X	Y	X^2	Y^2	XY
9	20	81	400	180
8	21	64	441	168
8	18	64	324	144
7	16	49	256	112
5	16	25	256	80
5	14	25	196	70
4	17	16	289	68
46	122	324	2,162	822

Since we have seven pairs of test scores, $n = 7$. Recall

$$r = \frac{n\Sigma XY - (\Sigma X)(\Sigma Y)}{\sqrt{n\Sigma X^2 - (\Sigma X)^2}\ \sqrt{n\Sigma Y^2 - (\Sigma Y)^2}}$$

Therefore,

$$r = \frac{7(822) - (46)(122)}{\sqrt{7(324) - 46^2}\ \sqrt{7(2{,}162) - 122^2}}$$

$$= \frac{5754 - 5612}{\sqrt{2{,}268 - 2{,}116}\ \sqrt{15{,}134 - 14{,}884}}$$

$$= \frac{142}{\sqrt{152}\ \sqrt{250}}$$

$$= \frac{142}{(12.33)(15.81)} = \frac{142}{194.9} = .73$$

Was your guess near .7?

CLASS EXERCISE

10-26 Estimate the value of r for the data in Figure 10-1. $r =$ ____________

10-27 Now compute the value of r.

X	Y	X^2	Y^2	XY
6	46			
7	47			
8	50			
8	51			
10	53			

10-28 Estimate the coefficient of correlation for the data in Figure 10-2. ________

10-29 Compute this coefficient of correlation. ________

X	Y	X^2	Y^2	XY
16	73			
16	71			
17	60			
18	66			
18	64			

THE SIGNIFICANCE OF r

Suppose a *sample* of pairs of data has a nonzero coefficient correlation. Can we be sure this reflects nonzero correlation in the entire population? As usual, the answer is: if our sample statistic r is far enough away from zero, then the population parameter is probably not zero.

If the two distributions that we are comparing are each approximately normal, then if we take many samples and compute many values of r, we will get a distribution of r's which has a known shape. This makes it possible for statisticians to compute critical values which cut off the ordinary 99 percent from the rare 1 percent of such sample outcomes. For example, if two variables in reality are uncorrelated, and we sample 15 pairs of data on these two variables, then when we compute r, which is the estimate of the coefficient of correlation from our sample data, we might not get zero. In fact according to the table of critical values, 99 percent of the time we would get a value of r between $-.64$ and $+.64$. A sample value of r between these two numbers is not unusual and should not be used to claim that we have discovered some nonzero correlation between these variables. On the other hand, a value of r greater than .64 would be extremely rare and would indicate that there was some positive correlation. A complete table of the critical values of r appears in Table B-6.

Example

Is there nonzero correlation between the length and the weight of mosquitos if a sample of 25 mosquitos led to a value of $r = .58$? Use $\alpha = .01$. You may assume that both the length and the weight of mosquitos are normally distributed. (In the remainder of this chapter this assumption is implied.)

Solution

We let the Greek letter for r, ρ (rho), stand for the true coefficient of correlation between mosquito length and weight.

$H_o : \rho = 0$; in words, mosquito length and weight are uncorrelated
$H_a : \rho \neq 0$
$n = 25$
$\alpha = .01$
$r_c = \pm .51$

Decision rule

I will reject the null hypothesis if my outcome is less than $-.51$ or greater than $.51$. Outcome: $r = .58$.

Conclusion

Since $.58 > .51$, I reject the null hypothesis. I am 99 percent sure that there is some positive correlation between the length and weight of mosquitos. This means that knowing a mosquito's length can be helpful in predicting its weight.

ASSIGNMENT

An old statistician named Fatterman
Drew for his grandma a scattergram
When the dots' disarray
Upset her he'd say
O, whatever is the matter, gram?

1 The following data were taken from a sample of Venusians where X is the finger span on right hand and Y is the finger span on middle hand.

Venusian	X, Inches	Y, Inches
XGL	14	6
LYR	3	7
BRL	8	5
DLG	9	8
EJZ	17	9

a Guess at the value of r. (*Hint:* A scattergram may help.)
Compute the following:
b ΣXY
c ΣX
d ΣY
e ΣX^2
f ΣY^2
g n
h r
i Test for nonzero correlation between X and Y in the entire population. Use $\alpha = .05$.

Directions: For Assignments 2, 3, and 4, repeat Questions a through i given in Assignment 1 for the following sets of data:

2

X = Income in Thousands of Dollars	5	9	11	14	14	20	25
Y = Amount Given to Charity	200	250	300	310	380	400	420

3

X = Age at Marriage	13	17	19	19	20	25	39
Y = Number of Engagements	1	1	3	1	4	5	1

4

X = Number of Children Under 18	1	1	2	4	4	5	5	5	7
Y = Number of Days Parents Vacationed Without Children Last Year	14	0	5	6	3	14	4	4	10

5 Is there any correlation between the number of children per household and the number of pets per household in Claremont? A random sample of 15 homes gave the following data:

X = Number of Children	0	3	1	5	0	2	6	1	5	2	3	1	0	3	0
Y = Number of Pets	0	1	0	0	1	1	3	1	1	0	2	0	1	1	1

a Compute the sample statistic r.

b Test for nonzero correlation in the population using the .01 significance level.

6 Does the following sample data show any correlation between the grade point average of graduating seniors at Unity University and their high school average at Happy High? Use .01 significance level.

High School Average X	Grade Point Average Y	High School Average X	Grade Point Average Y
72	1.9	84	3.1
73	2.0	85	2.8
74	2.1	88	3.0
75	2.0	89	3.2
75	2.1	90	3.4
76	2.3	93	3.1
77	2.2	94	3.5
79	2.5	96	3.5
80	2.7	97	3.6
81	2.9		

7 Agility tests of sit-ups and squat thrusts for 30 seconds were given in gym class to a random sample of 10 girls. Is there a correlation between the results on both tests? Test the significance of r using $\alpha = .01$.

Number	Sit-ups X	Squat Thrusts Y
1	40	28
2	36	30
3	24	22
4	42	28
5	30	24
6	35	26
7	21	18
8	34	29
9	34	22
10	32	30

8 Alvin and three of his friends compared their grades in the business management course and the business law course:

X = Business Management	95	80	85	85
Y = Business Law	76	65	60	75

a Compute the value of r.

b Does this indicate nonzero correlation in the whole class? Let $\alpha = .05$.

9 A young sports fan, David, collects baseball cards. He considered his cards to

be a random sample of baseball players. He investigated the relationship between the number of homeruns a player hits and the number of doubles he hits. Here are his data for 14 players who all hit at least 100 doubles by 1973. Is there significant correlation at the .05 significance level?

Player	Homers	Doubles
Tolan	66	127
Clarke	27	147
May	204	204
Pinson	246	453
Alou, F.	206	359
Powell	291	230
McCraw	67	133
Alou, M.	31	234
Alou, J.	28	150
Edwards	80	195
Petrocelli	185	191
Kranepool	85	157
Cater	61	184
Tovar	39	211

10 A sample of 12 insects captured in New Jersey after an atomic reactor accident gave data on the number of ears and the number of mouths of each insect. If $r = -.69$, what could be said about the population of all such insects? Let $\alpha = .05$.

11 A sample of evening students at Partime U. showed a coefficient correlation of .83 between the students' ages and their grades. Test the significance of r if $\alpha = .01$ and the sample consisted of 30 students.

12 IQ tests were given to 19 married couples from Pomona Flats. If r was computed to be $-.43$, what conclusions can you draw? Let $\alpha = .05$.

PREDICTION

Evy and Joe play a game as follows. Evy spins a spinner which selects a number from 1 to 20 at random. Then Joe spins the spinner in an attempt to spin a higher number than Evy did. If he succeeds, he wins a dollar, if not, he loses a dollar. For example, suppose the first five rounds were:

Round	Evy	Joe
1	9	7
2	13	1
3	8	17
4	11	11
5	6	9

TABLE 10-3

X = Value of First Spin	Y = Value of Joe's Average Win per Game
2	80
4	70
6	42
8	32
10	−06
12	−16
14	−34
16	−56
18	−84
20	−100

Joe lost rounds 1, 2, and 4 and won rounds 3 and 5, therefore, he is now out $1. In a given round, after Evy spins, Joe has some idea of the probability that he will win or lose that round. If she spins low, his chances are good; but if she spins high, then he will probably lose. To get some idea of the probabilities involved, we programmed a computer to play this game many times.

The computer played it 100 times when the first spin was a 2. Joe won 90 times and lost 10 times, so he won $80. Therefore, he won on the average 80 cents per game when the first spin was a 2. The computer then played 100 times when the first spin was a 4. Joe's average win was 70 cents. We let the computer continue in this way. The results appear in Table 10-3. Now if Evy spins a 9, what is the average amount that Joe will win? From our computer simulation it seems that the answer is somewhere between a gain of 32 cents and a loss of 6 cents. Let us draw a scattergram for this data (Figure 10-3).

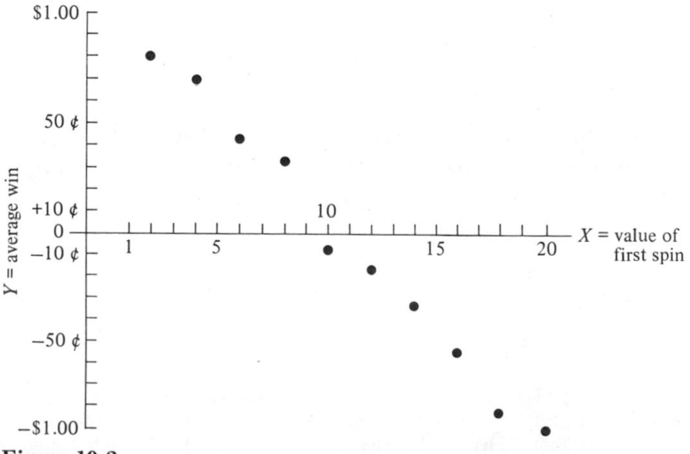

Figure 10-3

The dots in the scattergram are almost in a straight line. We can compute r and test for nonzero correlation. If there is indeed some linear correlation, we can find the so-called "best" line through these points and use it to estimate the value of Joe's average win when Evy spins a 9.

For this data, $r = -.996$ (see Assignment 13). For $n = 10$ and $\alpha = .01$, we have $r_c = \pm.76$, and thus we are confident that there is some* negative linear correlation between X and Y. If there is correlation in the population, then the formula for the best line is

$$\tilde{Y} = b(X - m_X) + m_Y$$

In this formula $\tilde{Y}$ stands for the predicted value of Y corresponding to a given value of X; m_X is the mean of the X's; m_Y is the mean of the Y's; and

$$b = \frac{n\Sigma XY - (\Sigma X)(\Sigma Y)}{n\Sigma X^2 - (\Sigma X)^2}$$

is called the **slope** of the line. In our problem $m_X = 11$, $m_Y = -7.2$, and $b = -10.3$ (see Assignment 13). Therefore,

$$\tilde{Y} = -10.3(X - 11) - 7.2$$

In our sample, Evy spun a 9, so $X = 9$ and

$$\tilde{Y} = -10.3(9 - 11) - 7.2 = 20.6 - 7.2 = 13.4$$

or about 13 cents. Therefore, when Evy spins a 9, Joe expects to win approximately 13 cents, that is, he should win more times than he loses.

The process of finding the best-fitting line and using it to make predictions is called **linear regression.** The line itself is sometimes called the **regression line.**

Another Example

Public Educational Service Testing (or PEST) has two versions of the same musical aptitude test. A sample of persons were given both form A and form B. Here are the test results:

A	47	22	10	80	4	50
B	37	11	0	72	6	30

* We have proved that there is some nonzero correlation. It may be very little indeed. The smaller it is, the less useful it is for making predictions. For a further treatment of this topic, see a textbook such as W. J. Dixon and F. J. Massey, Jr., "Introduction to Statistical Analysis," McGraw-Hill, New York, 1969.

If somebody scored a 30 on test A, what do you think that person will score on test B?

CLASS EXERCISE

10-30 Guess the answer to the above question. ______________________

If we compute r for the data above, we get $r = .97$. (See Assignment 14.) When $n = 6$ and $\alpha = .01$, $r_c = \pm .92$. Therefore, we have some linear correlation and the formula to predict a score on test B is

$$\tilde{B} = b(A - m_A) + m_B$$

where

$$b = .90$$
$$m_A = 35.5$$
$$m_B = 26$$

(See Assignment 14.) Therefore, the prediction formula is

$$\tilde{B} = .9\ (A - 35.5) + 26$$

If A = 30, we get

$$\tilde{B} = .9\ (-5.5) + 26$$
$$= -4.95 + 26$$
$$= 21.05$$

Chapter Summary

We have seen that r, the correlation coefficient, is a statistic which measures the linear relationship between two variables. If there is a significant correlation, then it may be useful to find the equation of the line which idealizes this linear relation. We can then use this line to predict which value of one variable corresponds to any given value of the other variable.

Words and Symbols You Should Know

Best-fitting line
Linear correlation
Linear regression
Negative correlation
Pearson correlation coefficient
Positive correlation
Prediction formula
Regression

Scattergram	r
Slope	r_c
Zero correlation	Y
b	ρ

ASSIGNMENT

I predict that your future with Lief
Is sure to bring nothing but grief
He drinks, he smokes
And he tells dirty jokes
And his money is stolen, the thief

13 Verify the values of r, m_X, m_Y, and b in the computer simulation in this chapter.

14 Verify the values of r, b, m_A, and m_B in the example following Exercise 10-30.

15 An experiment was designed in which people were told that natives on a Pacific island used the length of the chief's left hand as a measure of length. The people were then shown three line segments. The first was 1 hand long. They were then asked to estimate the lengths of the other two. Five persons made the following estimates:

Person	X = First Length	Y = Second Length
Alice	4 hands	2 hands
Barbara	2.5 hands	1 hand
Cathy	4.5 hands	2 hands
Donna	3 hands	1.5 hands
Edna	4 hands	2 hands

a Compute the value of the slope b.

b Write the prediction formula.

c If Fran estimates the first length to be 3.5 hands, predict what she will estimate as the second length.

d Compute r and test for nonzero correlation, with $\alpha = .05$.

e Comment on the prediction in part c.

16 In an experiment on communication among honey bees the following data were recorded.

Bee Number	Length of Time (in Seconds) Bee Performed "Tail Wagging" upon Return to Hive	Distance from Hive to Feeding Place (in Meters)
1	0.5	200
2	0.5	300
3	1.0	500
4	1.4	1,000
5	2.1	2,000
6	3.2	3,500
7	4.0	4,500

a Show that there is significant correlation between time spent tail wagging and distance to the food. Use $\alpha = .01$.
b Predict the distance to the food if a bee shows up at the hive and tail wags for 3 seconds.

17 Using the data in Assignment 1,
a Compute b.
b Write the prediction formula.
c Predict the second variable if $X = 10$ inches.

18 Using the data in Assignment 2, predict Y when $X = \$10,000$.
19 Using the data in Assignment 3, predict Y if $X = 19$.
20 Using the data in Assignment 4, predict Y if $X = 3$.
21 Using the data in Assignment 5, predict Y if $X = 4$.
22 Using the data in Assignment 6, predict Y if $X = 82$.
23 Using the data from Assignment 7, predict Y if $X = 34$.
24 Using the data from Assignment 8, predict Y if $X = 90$.
25 a Design your own correlation and prediction experiment.
b After your instructor has approved your design, carry out your experiment.
26 Make up your own question and use the data in Appendix A to answer it.
27 Read The Dynamics of a Heroin Addiction Epidemic, *Science*, pp. 716–720, August 24, 1973 for an example of a study where results were analyzed for significance using a t test, a chi-square test, and correlation.

Readings from J. Tanur (ed.), "Statistics: A Guide to the Unknown," Holden-Day, San Francisco, 1972.

28 Registration and Voting, Tufte, p. 153. If you know the percentage of people in a city who are registered to vote, how well can you predict the percentage who will actually vote?
29 Statistics, the Sun and the Stars, Whitney, p. 385. What can be learned about the sun by methods of correlation?
30 Adverbs Multiply Adjectives, Cliff, p. 176. Using linear regression to estimate ratings for the intensity of adjectives.
31 Preliminary Evaluation of a New Food Product, Street and Carroll, p. 220. Using linear regression to predict the effects of a particular diet.

Readings from The New York Times

32 Read about a link between a precancer condition and users of a birth control pill on page 84, September 27, 1970.

Review

33 A student investigating smoking habits on campus wanted to see if there was a relationship between the sex of the smoker and smoking habits. In a sample of 54 males, she found 6 smoking cigarettes only, 16 smoke something else only, 26 smoke both, and 6 smoke neither. In a sample of 46 females, she found 7 smoke cigarettes only, 11 smoke something else only, 19 smoke both, and 9 smoke neither. What are the results of an analysis of this data using $\alpha = .05$?
34 A newspaper reported that, based on its poll, 63 percent of local residents favored allowing any store to open for business on Sundays. A student inter-

viewed 200 local residents and 154 were in favor of allowing stores to open. Using $\alpha = .01$ does this tend to confirm or deny the newspaper's figure?

35 A student read an article in *The New York Times* that said of 1,057 homosexual men sampled, 86 percent said that they would not want psychiatric treatment for homosexuality and 14 percent said that they would. This sample represented the entire U.S. male homosexual population. A random sample of 100 students at a local Gay Liberation meeting showed that 39 percent would like psychiatric help and 61 percent would not. Using $\alpha = .01$, is there a significant difference between the two populations sampled?

36 A telephone company public relations statement claims that the average length of time waiting for a dial tone is 2.5 seconds. A random sample of 49 waiting times had a mean of 3.4 seconds and s_{pop} of 2.1 seconds. Test the telephone company's claim at the .05 significance level.

37 Do Venusian rocketships average the same distance per year as Saturnian rocketships? A random sample of 50 Venusian rocketships showed an average of 186,000 miles per year with a standard deviation of 40,000 miles, while 20 rocketships from Saturn averaged 150,000 miles with $s = 50{,}000$ miles. Use $\alpha = .05$ and see if the difference is statistically significant. (*Note:* These figures are fictitious since the true figures are classified.)

38 "The Hollywood Squares" television show claimed that 48 percent of Americans believe that the devil exists. Terry decided to estimate what percentage of the college staff believed in the existence of the devil. He interviewed 200 persons and 92 affirmed that they believed in the devil.

a Estimate what percentage of the staff believe in the devil.

b Find the 99 percent confidence interval for their estimate.

ELEVEN

Distribution-Free Tests

Rank has its privileges.

Anonymous

Most of the statistical tests that we have talked about in this book have had something to do with population **parameters.** For example, we asked, "Is the *mean* of population A equal to the *mean* of population B?" This is called a **parametric hypothesis**, and such a test is called a **parametric test.**

You may not have paid much attention to it, but many of the parametric tests make certain assumptions about the populations involved. For example, when we took small samples from a population to test a hypothesis about the mean of the population, we said that the test procedure was valid if the original population was *normally* distributed. If we have no good reason to believe that the population *is* normally distributed, then our test procedure will not work because the tabled critical values will not apply. Often, however, we want to be able to test some hypothesis in spite of the fact that we do not know enough about the population to be able to make such assumptions. Then it is necessary to find a procedure that does *not* depend on the distribution of the population. Or to put it another way, we want a procedure that will work on *any* population. Such a procedure is called **distribution free.**

Usually a distribution-free test is not as efficient as a parametric test. This means that to get conclusive results with a distribution-free test you need larger samples than for a corresponding parametric test. That makes sense because in the parametric case you already know something important about the population before you begin.

It is typical in a distribution-free procedure to ask general questions like, "Is it likely that these two samples come from the same population?" rather than questions about a particular parameter such as, "Is it likely that the *mean* of this population is equal to 0?" For these reasons distribution-free procedures are commonly called **nonparametric** procedures. In many discussions the two terms, distribution-free and nonparametric are equivalent.

Another example of a situation where we might want to use a nonparametric hypothesis test occurs when items in the population are ranked rather than measured (as in tests about means) or counted (as in tests about proportions). All of the tests and procedures that we have discussed so far which depend on measurement have assumed that the population *can* be measured in a meaningful way, and more than that, that our measurements were of a kind where it made sense to use the *mean* as the best kind of average. Not all studies lead to measurements like this. A common occurrence in statistical studies is one which involves a **ranking scale.**

Here is an example. A group of six paintings is shown to some professional artists from an Asian culture. They are asked to rank them in order from their most favorite (number 1) to their least favorite (number 6). Then a group of professional artists from a Western culture is given the same task. We are trying to find out if cultural background affects artistic taste. Technically, we are testing the hypothesis: "The rankings given by Western experts have the same distribution as the rankings given by Asian experts." We would do this by somehow getting an overall ranking by the Western experts and an overall ranking by the Asian experts. One way to do this might be to add the rankings for picture *A*, *B*, etc., and give the picture with the lowest sum rank 1. Then we would have to devise some way to decide if the two sets of rankings from these two samples of experts were different enough to be significant.

Before we get to that problem, let us just examine some rank measurements to see how they differ from the measurements we have studied so far. For instance, when one of the experts rates the paintings, we would see something like this:

Painting	*A*	*B*	*C*	*D*	*E*	*F*
Rank	3	1	5	2	4	6

These ranks are not like the measurements we have studied before for several reasons. For instance, we do not assume that a score of 4 is twice as bad as a score of 2. We do not know for paintings marked 1, 2, and 3 if the artist thought they were all close in value or not. It is possible that he liked 1 a *little* more than 2, but he liked 2 a *lot* more than 3. In other words, a difference of 1 does not have a consistent meaning. All we know is the *order* in which the paintings are ranked. In this situation we can use a type of nonparametric test called a **rank test.**

Sometimes we use a rank test because we *have* to: because the measurements from the very beginning are rank measurements, or the populations involved are not normal. Sometimes we use a rank test because it is *convenient*. For example, suppose we are comparing two types of paper airplanes to see which type stays in the air longer. If we do not care exactly *how* long they stay in the air but just which type is better, then we can convert the times to ranks. Here is an illustration:

Plane Number	Type	Seconds in Air	Rank
1	*A*	10.1	1
2	*B*	8.5	4
3	*A*	9.5	3
4	*B*	8.2	5
5	*A*	9.7	2
6	*B*	8.0	6

Now we can state a statistical hypothesis about the ranks, and use a nonparametric procedure for testing the hypothesis. For example, we might have H_0: The ranks for both types of planes are the same. (Essentially, this would mean that we expect both types of planes to perform equally well. We would expect them both to have the same share of first place ranks, second place ranks, etc., if we kept on testing planes.)

It is helpful to realize that once we have converted all the air times to ranks, we have "thrown away" some information about the original distributions of air times. We are now only taking into account the *order* in which the planes finished, not for instance, whether the second place finisher was *close* to the first place finisher. We are no longer basing our test, for example, on the fact that we originally had some idea about the *average* flying time for type *A* planes. Since we are ignoring information about the original population distributions when we use rank scores, you can see that rank tests are one example of distribution-free tests.

CLASS EXERCISE

In Figure 11-1 we have eight photographs of male faces.

11-1 Rank them in order from most pleasing (1) to least pleasing (8), according to your own taste.

1 ________ 2 ________ 3 ________ 4 ________

5 ________ 6 ________ 7 ________ 8 ________

11-2 Work out some system for pooling the information to come up with a ranking which represents the view of the male students in the class. Do the same for females.

MALE CONSENSUS

1 ________ 2 ________ 3 ________ 4 ________

5 ________ 6 ________ 7 ________ 8 ________

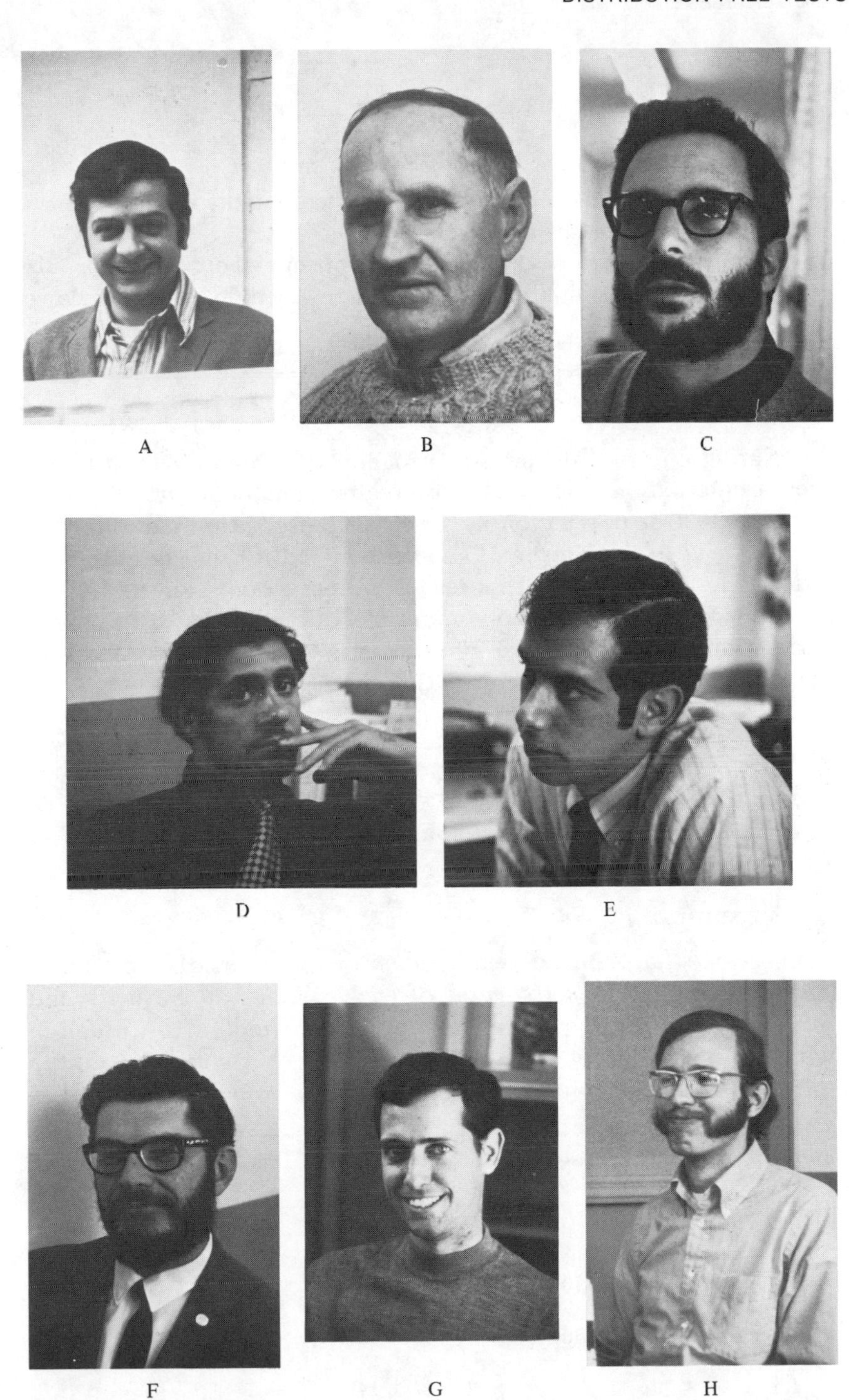

Figure 11-1

FEMALE CONSENSUS

1 ________ 2 ________ 3 ________ 4 ________

5 ________ 6 ________ 7 ________ 8 ________

11-3 Discuss methods of using the data from your class to see if the male and female students basically agree on the rankings.

11-4 If you assume that your class is a random sample of your whole college population, do you think you could tell if there would be any difference in rankings between males and females campuswide? ________________

We are going to illustrate *two* procedures which deal with rank measurements. The first is a procedure analogous to the two-sample t test of population means. This procedure can often be used instead of the t test and should be used if the original populations are not normal. This test is called a **rank sum test.** The second procedure is a nonparametric equivalent to the Pearson correlation coefficient procedure, but it is based on rank measurement and is called a **rank correlation test.**

TWO-SAMPLE RANK COMPARISON: A RANK SUM TEST

First Example

Suppose we have raised two species of racing snails, species A and species B. We set several of each species at the mark and release them. Suppose we are only interested in the *order* in which they finish, not in their exact times. We would record the data for the snails as follows:

Finishing Place	Snail Type	Finishing Place	Snail Type	Finishing Place	Snail Type	Finishing Place	Snail Type
1st	A	6th	A	11th	A	16th	B
2d	A	7th	A	12th	B	17th	B
3d	A	8th	A	13th	B	18th	B
4th	A	9th	B	14th	A	19th	B
5th	B	10th	B	15th	A		

CLASS EXERCISE

11-5 Does this sample indicate that species *A* snails are the better racers?

____

11-6 How is this reflected in the rank scores? _______________

A statistician, F. Wilcoxon, developed a procedure for comparing the rankings in two populations. We will explain that procedure using the snail data. The procedure is often called the **Wilcoxon test.** Think of the finishing order of the snails as their ranks. Rearrange the data as shown below, putting finish place under species. Then add the ranks in each column to get the rank sum.

Species *A*	Species *B*
1	5
2	9
3	10
4	12
6	13
7	16
8	17
11	18
14	19
15	
71	119
$n_A = 10$	$n_B = 9$

Since the sample of species *A* snails did so much better than the sample of species *B* snails, *A* has most of the low ranks. If the species were about equal in racing ability, these samples would probably have about the same ranks. We would expect to see high and low ranks sprinkled throughout both columns. So Wilcoxon's procedure basically tests whether the difference between the two rank sums is significantly different from what we would expect if the two populations were equal. His procedure has to take account of the fact that the two samples might not be the same size, so it is not quite as simple as just looking at the two rank sums.

The computation is designed to examine the ranks in the *smaller sized* sample. (If both samples are the same size, then you choose either one.) In this case species *B* had a smaller size sample ($n_B = 9$). We look at the rank sum for species *B*. It is 119.

We have to decide if it is unusual enough to be considered significant. We have to compare it with the sum predicted by the null hypothesis. The null hypothesis says that both species have equal racing ability. (The alternative hypothesis says they do *not* have equal ability.)

We need a way to predict what the rank sum would be if the two populations of snails were of exactly equal ability. In order to do this, we have to discuss how to record a tie score if several snails are equal. The standard way is to split the rankings equally among any snails which tie. For example, suppose the first three snails are tied. Instead of giving them ranks 1, 2, and 3, we reason that the first three places are worth a total of 6 points $(1 + 2 + 3 = 6)$. We split this among the three snails so that each gets two points $(6/3 = 2)$. As another example, suppose that two snails are tied for fifth place. Instead of giving them both a rank of 5, we reason that they are the fifth and sixth snails and are worth 11 points $(5 + 6 = 11)$. We split this between them so that each gets 5.5 points $(11/2 = 5.5)$. The basic idea is that this method of scoring ties always produces the *same rank sum,* no matter how the ranks are split.

Here is a sample illustration. Suppose there are three snails: Bob, Gene, and Henry. In a race of three snails with no ties, we will get the ranks 1, 2, and 3 for a total of 6.

Bob	1	
Gene	2	
Henry	3	
	6	(rank sum)

Now, suppose Bob and Gene tie for first.

Bob	1.5	
Gene	1.5	$\left(1.5 = \frac{1+2}{2}\right)$
Henry	3	
	6	(rank sum)

Or suppose that Gene and Henry tie for second.

Bob	1	
Gene	2.5	$\left(2.5 = \frac{2+3}{2}\right)$
Henry	2.5	
	6	(rank sum)

Or suppose that they all tie for first.

Bob	2	$\left(2 = \frac{1+2+3}{3}\right)$
Gene	2	
Henry	2	
	6	(rank sum)

You see that this system of scoring ties always gives the same rank sum.

CLASS EXERCISE

11-7 How would you rank five people if three are tied for first place.

Finish	Name	Rank
Tied for first	Joe	________
	Jim	________
	Mike	________
Next	Marty	________
Last	Fenster	________

What is the rank sum? ________________________________

What is the *average* rank? ________________________________

11-8 Suppose they finished with two tied for last place?

Finish	Name	Rank
First	Joe	________
Second	Jim	________
Third	Fenster	________
Tied for fourth	Mike	________
	Marty	________

What is the rank sum? ________________________________

What is the *average* rank? ________________________________

11-9 Suppose they all tied for first?

Finish	Name	Rank
All tied	Joe	________
	Fenster	________
	Jim	________
	Marty	________
	Mike	________

What is the rank sum? ______________________________

What is the *average* rank? ______________________________

In our original example about the snails, there were 19 snails running. The sum of $1 + 2 + 3 + \cdot \cdot \cdot + 19$ is 190. Therefore, after all 19 are ranked, the rank sum will be 190, and the average rank will be 10. $(190/19 = 10)$

Therefore, if the null hypothesis is true and species *A* snails are equal to species *B* snails, then the average rank for an *A* snail is 10 and the average rank for a *B* snail will be 10. Since there are nine *B* snails altogether, we expect species *B* to have a rank sum of $9 \times 10 = 90$. We expect species *A* to have a rank sum of $10 \times 10 = 100$.

Now, in our sample, if you look back at the rankings, the sum for species *B* was not 90 but rather 119. What is the probability of that happening if the null hypothesis is true? This probability can be calculated exactly by the methods of

$$p = \frac{\text{number of favorable outcomes}}{\text{total number of outcomes}}$$

but it is pretty complicated because there are so many different possible ways for the snails to finish. Tables of possible results and their probabilities given that H_0 is true have been published. We have reproduced in Table 11-1 a table needed for problems where both sample sizes are between 5 and 10. For larger samples there is another method using critical scores from a normal curve table which we will illustrate shortly. We will use the letter *W* (for Wilcoxon) to represent the rank sum of the *smaller* size sample.

Table for Rank Sum Test

The values in Table 11-1 are the critical values for *W*, the rank sum of the smaller size sample.

In our problem about the snails, the sample sizes were 9 and 10. You can see that if we choose $\alpha = .05$, then the critical values for *W* are 65 and 115. This means that if the null hypothesis were true, then 95 percent of experimental results would have a rank sum between 65 and 115. Our experimental value was $W = 119$. Therefore, we have statistical evidence at $\alpha = .05$ that species *A* snails are faster than species *B* snails.

TABLE 11-1

Sample Sizes Smaller, Larger	W_c	
	$\alpha =$ 01	$\alpha = .05$
5, 5	15, 40	17, 38
5, 6	16, 44	18, 42
5, 7	17, 48	20, 45
5, 8	18, 52	21, 49
5, 9	18, 57	22, 53
5, 10	19, 61	23, 57
6, 6	23, 55	26, 52
6, 7	24, 60	27, 57
6, 8	25, 65	29, 61
6, 9	26, 70	31, 65
6, 10	28, 74	32, 70
7, 7	32, 73	36, 69
7, 8	34, 78	38, 74
7, 9	35, 84	40, 79
7, 10	37, 89	42, 84
8, 8	44, 92	49, 87
8, 9	45, 99	51, 93
8, 10	47, 105	53, 99
9, 9	57, 114	63, 108
9, 10	59, 121	65, 115
10, 10	71, 139	78, 132

Summary of Rank Sum Test

1 Identify populations.
2 State null hypothesis.
3 Choose significance level.
4 Observe samples, rank the data.
5 State decision rule.
6 Find *W*, the rank sum for the smaller sample. This is your experimental outcome.
7 State conclusion.

COMPUTER SIMULATION

You might like to see the results of a computer simulation designed to show that the critical scores for *W* in the table are reasonable. We put into the computer two populations which have equal distributions of ranks. Imagine that we put in a large population of type *A* snails and a large population of *B* snails which were of *equal* racing ability. Then we have the computer printout a possible finishing order for a race between 10 *A* snails and 9 *B* snails. To simplify things, we assumed no snails would tie. Since the snails are of equal ability, an *A* snail is just as likely as a *B* snail to finish in any particular spot from 1 to 19.

We then have the computer examine the order in which they finish, and compute *W* for that order. Then we repeat this whole procedure 100 times to get a list of typical *W*'s. This will give us some idea of what values of *W* are common when the two populations are equal.

TABLE 11-2

66	66	68	69	70	70	72	72	72	73	74	74	75	76	76	78	78
79	80	81	81	81	81	81	81	81	82	82	82	82	83	84	84	84
85	85	85	86	87	87	87	87	87	88	88	88	88	88	88	89	90
90	90	90	91	91	91	92	92	92	93	93	93	94	94	94	95	96
97	97	99	99	99	99	100	100	101	101	101	101	102	102	103	104	104
104	105	106	109	110	111	112	113	113	113	116	117	124	129			

Table 11-2 shows the 100 values of W that the computer generated. They are rearranged in numerical order. If you look at the list of W's, you can see that 96 of them are between 65 and 115. That should make the critical values of 65 and 115 seem reasonable. A sample value of W less than 65 or more than 115 is unusual enough to make you suspect that the two populations are probably not equal.

Second Example

A high Priestess of Voodoo has developed two love potions which she wishes to compare. She selects 18 subjects at random and gives half of them potion A and half of them potion B. Then she watches and ranks the subjects from least affected (1) to most affected (18). The results are given in Table 11-3.

Procedure

1 Population 1: Males given type A potion.
  Population 2: Males given type B potion.

TABLE 11-3

Subject	Type of Potion He Got	Ranking
Abe	A	4
George	A	10
Jim	A	3
Joe	A	12
Larry	A	6
Carmine	A	1
Aaron	A	11
Mauro	A	5
Dennis	A	2
Ken	B	16
Bill	B	17
Doug	B	9
Frank	B	15
Mike	B	7
Arnie	B	18
Eliot	B	14
Tom	B	13
Ed	B	8

2 H_0 : Effect of potion A is same as effect of potion B.
H_a : Effect of potion A is not same as effect of potion B.
3 Choose $\alpha = .05$.
4 Observe the data above.
5 *Decision rule:* If the rank sum for type A potion is less than 63 or more than 108, we will have reason to reject H_0. (*Note:* Since n_A and n_B are both 9, we have arbitrarily chosen A as the sample to investigate. W is the rank sum for sample A.)
6 Compute W.

Potion A
4
10
3
12
6
1
11
5
2
$W = 54$

7 *Conclusion:* Since W is smaller than 63, we see that potion A has more low ranks than expected if H_0 were true. Therefore, we reject H_0. We have reason to believe that the potions are not equally effective. Evidently potion B is more effective than potion A.

CLASS EXERCISE

These next exercises are designed to show you that a formula which appears in the next section is a reasonable one. Consider two samples with $n_1 = 3$ and $n_2 = 5$.

11-10 There are $3 + 5 = 8$ rankings: 1, 2, 3, 4, 5, 6, 7, 8. Find their mean by adding them up and dividing by 8. ________________

11-11 The null hypothesis assumes that the mean of the rankings in the smaller sample is the same as the mean of the rankings in the larger sample and that both of these are equal to your answer to Exercise 11-10. Under this assumption, find the *sum* of the three ranks in the smaller sample. ____________

11-12 Does the formula

$$\mu_W = \frac{n_1(n_1 + n_2 + 1)}{2}$$

give the same answer as Exercise 11-11?

RANK TEST WITH LARGER SAMPLES

When the larger sample is bigger than 10, it turns out that the collection of *W*'s you would get in repeated sampling is distributed approximately normally, and you can find the critical *W* values by using the critical *z* scores from the normal table. We illustrate by doing the snail problem again. Because the sample sizes were 9 and 10, this would be a borderline case for using the normal curve. We will do it using the normal curve to show you that even in this case the critical values are close to the ones in Table 11-1. If you have larger samples, the normal approximation is even closer. We repeat the data.

Type *A* Snail	Type *B* Snail
1	5
2	9
3	10
4	12
6	13
7	16
8	17
11	18
14	19
15	
71	$W = 119$

The null hypothesis was: "Both populations of snails have the same distribution of ranks for racing."

Let n_1 stand for the size of the smaller sample. Let n_2 stand for the size of the larger sample. Now if the null hypothesis is true, then theory predicts that a distribution of sample *W*'s for the smaller sample would be distributed normally with

$$\mu_W = \frac{n_1\,(n_1 + n_2 + 1)}{2}$$

and

$$\sigma_W = \sqrt{\frac{n_1 n_2 (n_1 + n_2 + 1)}{12}}$$

We find W_c, the critical scores, by

$$W_c = \mu_W \pm z_c \sigma_W$$

In our problem we have $n_1 = 9$ and $n_2 = 10$. Therefore,

$$\mu_W = \frac{9(9 + 10 + 1)}{2} = \frac{9(20)}{2} = 90 \text{ for } B$$

(*Note:* This is the same value we had figured out following Exercise 11-10.)

$$\sigma_W = \sqrt{\frac{(9)(10)(9+10+1)}{12}} = \sqrt{\frac{(90)(20)}{12}} = \sqrt{150} = 12.25$$

Since $\alpha = .05$, we have $z_c = \pm 1.96$. Therefore,

$$W_c = 90 \pm 1.96\ (12.25) = 65.99,\ 114.01$$

You will notice that these critical values are close to the ones listed in the table we used before, 65 and 115.

Illustration

Here is another illustration where the values of n are large enough to use a normal approximation. We know that 30 students at Flink University signed up for calculus with Professor Hardhardt. She randomly split them into two groups of size 15. In class T she stressed *t*heory, and in class A she stressed *a*pplications. At the end of the year she gave the same test to both classes. Is there a significant difference, if $\alpha = .05$, in the results based on the data shown in Table 11-4?

Solution

1 Population 1: Students taught calculus by stressing theory.
Population 2: Students taught calculus by stressing applications.
2 H_0: The two methods will produce equal rankings on the test.

TABLE 11-4

Initials of Students	Class	Final Standing in Class	Initials of Students	Class	Final Standing in Class
NA	*T*	1	RB	*A*	16
HA	*A*	2	JC	*A*	17
DG	*A*	3	LA	*T*	18
SS	*T*	4	VB	*A*	19
MM	*T*	5	DB	*T*	20
WU	*A*	6	CS	*T*	21
CL	*A*	7	AW	*A*	22
TL	*T*	8	DF	*T*	23
AS	*A*	9	AB	*A*	24
CH	*A*	10	GS	*T*	25
PM	*A*	11	GT	*T*	26
RK	*T*	12	IK	*T*	27
WS	*A*	13	DT	*T*	28
ES	*A*	14	CP	*T*	29
JG	*T*	15	ZF	*A*	30

H_a: The two methods will produce different rankings on the test.

3 $\alpha = .05, \quad W_c = \mu_W + 1.96\,\sigma_W, \quad n_1 = 15,\ n_2 = 15$

$$\mu_W = \frac{(15)(15+15+1)}{2} = \frac{(15)(31)}{2} = 232.5$$

$$\sigma_W = \sqrt{\frac{(15)(15)(15+15+1)}{12}} = \sqrt{\frac{(15)(15)(31)}{12}} = \sqrt{581.25} = 24.11$$

Therefore,

$$W_c = 232.5 \pm 1.96\,(24.11) = 232.5 \pm 47.26 = 185.24, \text{ and } 279.76.$$

Decision rule: If W falls outside of the interval from 185.24 to 279.76, we will reject H_o.

4 *Compute* W: Since n_1 and n_2 are both 15, we can compute W by taking the rank sum of either class. We choose class T.

Class T Ranks
1
4
5
8
12
15
18
20
21
23
25
26
27
28
29
$W = 262$

5 *Conclusion:* The experimental value of W is *not* in the rejection region. We cannot state with confidence that one method is superior to the other for achieving good scores on Professor Hardhardt's test as far as the general population of calculus students is concerned.

6 *Comments:* It does appear intuitively that a large group of students taught theory ended up at the bottom of the heap. This suggests that another experiment should be done to try to identify what type of student that is. Also, one must keep in mind that the professor's test itself may be biased against students who learned primarily theory. One must ask, "What precisely was the professor *trying* to achieve?"

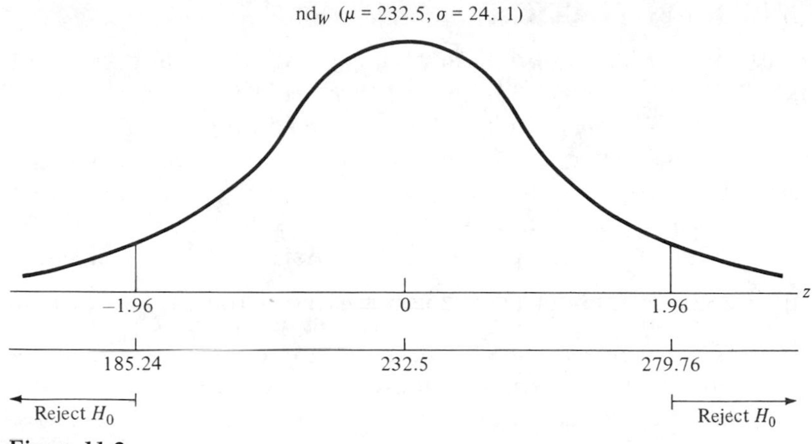

Figure 11-2

CLASS EXERCISE

11-13 Suppose you are trying to compare two methods of training water polo players. You split them into two groups randomly, put them through training, and then test them in some way. Suppose there are 12 athletes in group *A* and 18 in group *B*.

a *W* will stand for the rank sum for which group? ____________

b If $\alpha = .05$, what will the two values of W_c be? ____________

11-14 An entomologist raises pet spiders. She is experimenting with two types of diet. She randomly splits a new batch of 21 spider babies into two groups. She gives group 1 diet *A* and she gives group 2 diet *B*. After 3 weeks she measures the length of each spider's body. She ranks them from longest (1) to shortest (2). Here are her results:

Rank	Diet	Rank	Diet
1	*A*	11	*A*
2	*A*	12	*A*
3	*A*	13	*A*
4	*A*	14	*B*
5	*B*	15	*A*
6	*B*	16	*B*
7	*A*	17	*B*
8	*B*	18	*B*
9	*A*	19	*B*
10	*B*	20	*B*
		21	*B*

What is the value of *W* for these data? ____________

RANK CORRELATION

Let us reconsider the problem about the Western and Asian artists. This time we ask, "Is there any *correlation* in their taste?" For example, if they ranked the same paintings high and the same paintings low, we would get positive correlation. Whereas, if their tastes were opposite, we would expect negative correlation; that is, if we were given a painting and told that a Western artist rated it high, we could predict that an Asian artist would rate it low.

The procedure is similar to the one you already know for correlation. The computation was first suggested by Spearman, and the symbol for the statistic is usually r_s in his honor.

Suppose that we had asked several Western artists and several Asian artists to rank six paintings. And suppose the set of rankings in Table 11-5 is the set of rankings that each group agreed to.

TABLE 11-5

Painting	Asian	Western
A	1	5
B	3	3
C	2	6
D	5	1
E	4	4
F	6	2

If we draw a scattergram for this data, we get Figure 11-3. The scattergram of this sample of opinions indicates a somewhat negative correlation. To judge if it is strong enough for us to claim that there is also negative correlation between the populations, we need to compute a sample statistic and compare it with a tabled critical value. To compute the statistic:

1 List all the ranks in pairs. Let n equal the number of pairs. We

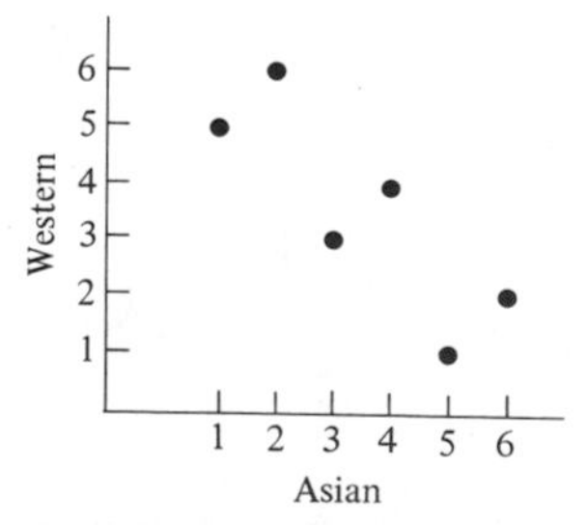

Figure 11-3

TABLE 11-6 ILLUSTRATION OF THE COMPUTATION OF r_s

Painting	Asian	Western	Difference	(Difference)^2
A	1	5	−4	16
B	3	3	0	0
C	2	6	−4	16
D	5	1	4	16
E	4	4	0	0
F	6	2	4	16
				$S = 64$

need at least six pairs for $\alpha = .05$, otherwise, the sample is too small to draw a conclusion.

2 Find the squares of the differences for each pair.
3 Find the sum of those squares and call it S.
4 Then

$$r_s = 1 - \frac{6S}{n(n^2 - 1)}$$

This is **Spearman's formula.** See Table 11-6.
Since there are six pairs, $n = 6$.

$$r_s = 1 - \frac{6(64)}{6(35)} = 1 - 1.828 = -.828$$

The idea would then be to compare this sample value of r_s with critical values from a table. The reasoning is the same as that for the correlation procedure we described in Chapter 10. The null hypothesis states that correlation in the population is 0. If the sample value of r_s is far enough from 0, we will reject the null hypothesis.

The table of critical values of r_s (Table 11-7) shows two values of r_s for each given number of pairs. It has been computed by probability methods that if your value of r_s is outside the two critical values for your number of pairs, then there is less than a .05 probability that the null hypothesis is true.

TABLE 11-7 CRITICAL VALUES OF r_s (RANK CORRELATION COEFFICIENT)

Number of Pairs n	$\alpha = .05$
6	−.77, +.83
7	−.71, +.75
8	−.69, +.71
9	−.67, +.68
10	−.62, +.64

For the example just worked out, the sample value of r_s was $-.828$. Looking at Table 11-7, we see for $\alpha = .05$ that r_s must be outside the interval $-.77$ to $+.83$ to reject the null hypothesis. Our sample value is outside this interval. Therefore, we have sufficient evidence at the .05 significance level to say that there is negative correlation between the populations represented by these artists. Evidently these populations have "opposite" artistic tastes.

If $n > 10$, you can find the critical values from this formula

$$r_c = \pm \frac{z_c}{\sqrt{n-1}}$$

Example

$n = 11$, $\alpha = .05$, then

$$r_c = \pm \frac{1.96}{\sqrt{11-1}} = \pm \frac{1.96}{\sqrt{10}} = \pm \frac{1.96}{3.16} = \pm .62$$

CLASS EXERCISE

11-15 If the ranks on two variables are exactly the same, then $r_s = 1$. Show that this is true based on this data.

Person	Height Rank	Weight Rank
Bailey	5	5
David	3	3
Hald	1	1
Seale	2	2
Thomas	4	4

11-16 If the ranks on two variables are exactly reversed and no ties occur, and then $r_s = -1$. Show that this is true based on this data.

Person	Sense of Humor Rank	Aggressiveness Rank
Zubrzycki	5	1
Yushkevitch	3	3
Woods	1	5
Vincze	2	4
Uhlmann	4	2

COMPUTER SIMULATION

To give you an idea of why the critical values of r_s are reasonable, we have set up a computer simulation. We stored in the computer a large population of "people." Each "person" was *ranked* for two things: his height and his yearly income. We arranged it so that for the population as a whole there was *no correlation* between these two items; that is, if we compute r_s for the *whole* population, we get zero.

Then we took a random sample of 20 of these "people" and computed a sample value of r_s. We then repeated this procedure 100 times to show a *distribution* of r_s's. This gives us an idea of what to expect when there is no correlation in a population. The results are shown in numerical order in Table 11-8.

TABLE 11-8 100 VALUES OF r_s COMPUTED FROM SAMPLES OF 20 PERSONS

−.674	−.471	−.450	−.448	−.391	−.389	−.367	−.313	−.280
−.257	−.248	−.245	−.233	−.226	−.224	−.214	−.197	−.189
−.188	−.165	−.155	−.155	−.153	−.152	−.149	−.146	−.141
−.114	−.113	−.111	−.100	−.099	−.093	−.090	−.084	−.080
−.077	−.074	−.071	−.065	−.050	−.045	−.044	−.041	−.024
−.017	−.002	.000	.005	.038	.044	.045	.047	.050
.056	.059	.065	.065	.069	.069	.075	.077	.092
.102	.104	.110	.113	.126	.134	.141	.143	.152
.156	.165	.174	.177	.197	.197	.198	.203	.205
.208	.211	.229	.247	.254	.268	.269	.286	.292
.305	.307	.320	.335	.343	.389	.402	.409	.412
.426								

You can see that almost all the values of r_s are between −.450 and +.450. This should make it reasonable to accept $\pm 1.96/\sqrt{19} = \pm .450$ as the critical value of r_s for 20 pairs of ranks. It would be very unusual to get a sample value of r_s outside this interval if there really was no correlation in the population. Or putting it the other way around, if you do get a sample value of r_s outside the critical values, you can be reasonably sure it happened because there is some correlation in the population.

Example

We close the chapter with another example of the rank correlation test. Let us investigate the correlation between the number of years of schooling a male completed and his yearly income 10 years after he leaves school. We will collect a random sample of 13 males from a particular community. Assign ranks as follows: least number of years of schooling is number 1, most number of years of schooling is number 13. Lowest income is number 1, highest is number 13. The data are in Table 11-9.

TABLE 11-9

Person	Years of Schooling	Income in Dollars
Babkin	12	7,000
Cantelli	13	7,800
Dodge	16	4,000
Esseen	12	8,000
Ferris	16	70,000
Gelfand	12	8,000
Hall	18	10,400
Ito	20	13,000
Katz	20	15,000
Madow	16	10,500
Pratt	14	15,000
Snell	14	14,000
Tippet	13	11,700

CLASS EXERCISE

11-17 Rank the years of schooling from 1 to 13.

Babkin ________ Ito ________

Cantelli ________ Katz ________

Dodge ________ Madow ________

Esseen ________ Pratt ________

Ferris ________ Snell ________

Gelfand ________ Tippet ________

Hall ________

11-18 Rank the incomes from 1 to 13.

Babkin ________ Ito ________

Cantelli ________ Katz ________

Dodge ________ Madow ________

Esseen ________ Pratt ________

Ferris ________ Snell ________

Gelfand ________ Tippet ________

Hall ________

Your answers should agree with the results in Table 11-10.

TABLE 11-10

Person	Schooling Rank	Income Rank	Person	Schooling Rank	Income Rank
Babkin	2	2	Ito	12.5	9
Cantelli	4.5	3	Katz	12.5	12
Dodge	9	1	Madow	9	7
Esseen	2	4.5	Pratt	6.5	11
Ferris	9	13	Snell	6.5	10
Gelfand	2	4.5	Tippet	4.5	8
Hall	11	6			

Solution

1 H_0: There is no correlation between years of schooling and income.
  H_a: There is correlation between years of schooling and income.

2 Let $\alpha = .05$. Our decision rule is found by:

$$r_c = \pm\frac{1.96}{\sqrt{n-1}} = \pm\frac{1.96}{\sqrt{12}} = \pm\frac{1.96}{3.46} = \pm.566$$

If our sample value of r_s is less than $-.566$ or more than $+566$, then reject H_0.

3 Now compute r_s for the same data. First we compute S.

Person	Difference of Ranks	Squares
Babkin	0	0
Cantelli	1.5	2.25
Dodge	8	64.00
Esseen	−2.5	6.25
Ferris	−4	16.00
Gelfand	−2.5	6.25
Hall	5	25.00
Ito	3.5	12.25
Katz	0.5	0.25
Madow	2	4.00
Pratt	−4.5	20.25
Snell	−3.5	12.25
Tippet	−3.5	12.25
		S = 181.00

$$r_s = 1 - \frac{6S}{n(n^2-1)} = 1 - \frac{6(181)}{13(168)} = 1 - \frac{1086}{2184} = 1 - .497 = .503$$

4 *Conclusion:* The correlation in the sample is .503. It is positive; but to be significant we have seen that it would have to be greater than $r_c = .566$. We do not have convincing evidence that years of schooling and income are correlated.

Chapter Summary

We have discussed two frequently used distribution-free tests: a rank sum test for equality of distributions and a rank correlation test for linear correlation. We have seen that these tests can be used when little is known about the populations being studied and that, therefore, they are very convenient but not as efficient as some other tests we have learned.

Words and Symbols You Should Know

Distribution free	r_s
Nonparametric	S
Rank correlation	W
Rank sum	W_c
Rank test	μ_W
r_c	σ_W

ASSIGNMENT

A major, because of his rank
Rode up front in a tank
The enlisted men walked
Behind him and talked
While his tank crossed a bank and sank

1 Repeat Exercises 11-10 to 11-12 with $n_1 = 5$ and $n_2 = 7$.

2 Repeat Exercises 11-10 to 11-12 with $n_1 = n_2 = 6$.

3 A waiter at Mount Cupid Honeymoon Lodge kept tabs on the number of meals missed during 1 week by newly married couples. Of the couples that he served during the first week in June he noted the following: The numbers of meals missed by couples on the first week of their honeymoon were 7, 5, 10, 8, 7, 10, and 17; while the numbers of meals missed by couples on the second week of their honeymoon were 0, 13, 3, 8, 10, 7, 10, and 6. Rank the 15 numbers above from 1 to 15, and by using the rank sum test, test the hypothesis that there is no difference between the number of meals missed by first- and second-week honeymooners. Use $\alpha = .05$.

4 The two foremost Frisbee experts in the nation are summoned to judge a group of contestants in the Eighth Annual Intercollegiate Frisbee Style Competition. The eight finalists are ranked by each judge. Would you say that the two judges have applied similar standards if these are their results? Test at $\alpha = .05$.

Contestant	Judge *A*	Judge *B*
Leonard	6	6
Damon	5	4
Harry	4	5
Isadore	7	8
Butch	8	7
Mary	1	1
Scoodles	3	2
Louise	2	3

5 Espionage agents in two different countries are trained to endure pain. Then at the Fourth Annual International Espionage Agent Pain Endurance Competition, nine agents from each country enter the contest. They are ranked from the one who can endure the *most* pain measured in volts (rank number 1) to the one who can endure the *least* (rank number 18). Here are the results:

Name of Agent	Country	Volts Endured
001	*X*	100
002	*X*	143
003	*X*	128
004	*X*	118
005	*X*	89
006	*X*	118
007	*X*	132
008	*X*	107
009	*X*	141
001	Z	130
002	Z	135
003	Z	142
004	Z	120
005	Z	60
006	Z	95
007	Z	120
008	Z	60
009	Z	143

Does this indicate at the .05 significance level that either country produces a superior agent? Convert the "Volts Endured" to ranks, then apply the rank sum test to compare the two brands of spies.

6 A random sample of 6 widowers and a random sample of 5 divorced men were given a test to measure how listening to romantic poetry affected their ability to

reason logically. Their scores were

Widowers: 4.1, 2.0, 8.2, 1.1, 6.0, and 0.2
Divorced men: 8.1, 1.0, 4.0, 5.8, and 1.9

a Show that using the rank sum test we cannot show that these two samples came from different populations. Let $\alpha = .05$.

b If we know that the original populations are normally distributed, we can do a two-sample t test on the difference between the sample means. Will the conclusion be any different from that of part a?

7 Two treatments are available for a certain illness. At a hospital 12 people enter with this illness: 6 are given treatment A and 6 are given treatment B. Shown in the table are the recovery times for each of the 12 patients. Convert the recovery times to ranks and use the rank sum test to show that the treatments are not equally effective. Let $\alpha = .05$.

Patient	Days to Recover	Type of Treatment	Patient	Days to Recover	Type of Treatment
1	5	*A*	7	8	*A*
2	8	*B*	8	9	*B*
3	6	*A*	9	6	*A*
4	7	*A*	10	10	*B*
5	8	*B*	11	9	*B*
6	5	*A*	12	9	*B*

8 A social psychologist was testing the ranking procedures used in a certain industry. She asked a foreman, Noah Fens, to rank each of the 15 men in his unit. She recorded both the rank and the race of each of the 15 men. Apply the rank sum test to the data and test for significance at $\alpha = .05$.

Worker	Worker's Race	Job Performance Rating
1	White	13
2	White	11
3	Black	7
4	Black	6
5	Black	10
6	White	12
7	Black	5
8	White	14
9	Black	1
10	White	15
11	Black	2
12	White	9
13	White	8
14	Black	4
15	Black	3

Can you give two different interpretations of these results? What further experimentation could you do to decide which interpretation is more likely to be true?

9 a Two samples of five dozen oranges each were ranked for quantity of juice. The oranges in the two samples had been grown from trees which received two different kinds of fertilizer. In one sample the rank sum was 3160, while in the second sample it was 4100. Does this indicate any difference at the .01 significance level?

b If the distributions are known to be normal, and the original samples had means of 3.25 ounces and 4.90 ounces while their standard deviations were about 1 ounce each, we could do a two-sample test on the difference between the sample means. Would this change your conclusion about the quantity of juice from the two populations of oranges?

10 A secretary in the English department, Andrew Blucher, says that the male faculty members tend to move the office thermostat down, while the female faculty members tend to move it up. Everytime the office is empty he sets the thermostat at 70°. Whenever someone adjusts it, he records the change in the setting. Here are his results. For 5 female members he recorded: +5, +2, −3, +10, and +4. For 6 male members he recorded: −3, +2, −1, −8, +2, and −5.

a Does this data support Andrew's contention? Use a rank sum test with $\alpha = .05$.

b Assume that the two distributions of thermostat changes are normal, and perform a two-sample test of sample means. Use $\alpha = .05$.

11 A professor thought there was some correlation between a student's grade and the row in which the student sat in class. It seemed to him that the students who sat closest got the highest grades. He collected data on 20 students.

Student	Row	Grade	Student	Row	Grade
1	4	C	11	2	A
2	4	B	12	5	C
3	1	A	13	3	C
4	1	B	14	5	B
5	4	A	15	1	A
6	5	C	16	2	B
7	2	A	17	4	D
8	2	B	18	4	C
9	1	A	19	3	B
10	5	D	20	3	C

Convert this data to rank scores, and test the hypothesis that there is no correlation between the two variables. Let $\alpha = .05$.

Hint: Let the closest rows have the lowest ranks, and let the highest grades have the lowest ranks. For example, the row rank for students 3, 4, 9, and 15 would be

$$\frac{1 + 2 + 3 + 4}{4} = 2.5$$

The grade rank for students 3, 5, 7, 9, 11, and 15 would be

$$\frac{1+2+3+4+5+6}{6} = 3.5$$

12 A product testing laboratory tested a random sample of the currently available brands of solid-state electronic chalkboard erasers. They ranked them according to how clean they left the board. (Because all brands could erase a 4-foot by 8-foot board in approximately 6 milliseconds, it was decided that speed of operation was probably not a deciding factor in choosing a brand.) Here are their ratings.

Brand	Price	Rating
A	\$1,000	Extra superduper
B	\$2,000	Superduper
C	\$3,000	Great
D	\$4,000	Good enough
E	\$5,000	Fair

Convert the prices and ratings to rank measures and show that there is perfect negative correlation between price and rating. Let the lowest price have rank 1, and let the lowest rating have rank 1. (What do you think the result would have been if you had let the lowest price have rank 1, but the *highest* rating have rank 1?)

13 A professor of physical education was investigating the effect of muscle structure on athletes' performance. He measured a property of calf muscle associated with elasticity to see if it correlated with high-jump performance. He tested 20 physical education majors. He ranked them according to muscle elasticity as determined by a laboratory test and also by high-jump skill as determined by competition. (Number 1 rank means most elastic muscle or best high jumper.)

Athlete	Elasticity of Muscle (Rank)	Finishing Place in High Jump	Athlete	Elasticity of Muscle (Rank)	Finishing Place in High Jump
1	1	7	11	11	1
2	2	2	12	11	9
3	3	12	13	13	14
4	4	10	14	14	11
5	5	17	15	15.5	3
6	7	13	16	15.5	5
7	7	8	17	17	4
8	7	19	18	18	20
9	9	15	19	19	16
10	11	6	20	20	18

Test at $\alpha = .05$ the hypothesis that muscle elasticity and high-jump skill are correlated in the population of athletes as represented by physical education majors.

14 A biologist was experimenting to see the effect of various levels of a naturally oc-

curring chemical on the average size of a certain type of microscopic animal which lives in fresh water ponds. He started 10 cultures with a small number of similar animals in each, then after a week he measured 100 animals from each culture and recorded their average size. He ranked the cultures from number 1 (containing highest amount of chemical) to number 10 (least amount). He ranked size from number 1 (biggest) to number 10 (smallest).

Experiment	Chemical Rank	Rank for Average Size of Animal
1	1	10
2	2	9
3	3	8
4	4	1
5	5	2
6	6	3
7	7	4
8	8	5
9	9	6
10	10	7

a Test at $\alpha = .05$ for significant correlation.

b Test again but only on experiments 4 through 10. Can you explain what might be happening?

15 A manufacturer of a cream for treating acne was trying to decide on a package design for a new product. The advertising department made up eight different designs and ranked them according to their past experience from number 1 (most masculine) to number 8 (least masculine). Then they test marketed a limited amount of the cream in each type of package by going to high school campuses and letting male students take one free sample of their own choice from among the display of all eight packages. At the end of the test-marketing period, they ranked the packages from number 1 (most popular) to number 8 (least popular) in an attempt to see if "masculinity" was an important aspect of buying psychology for this product. Test the data at $\alpha = .05$ for significant correlation between "masculinity" and popularity.

Package	Masculinity Rank by Advertising Department	Number of Samples Taken
1	1	840
2	2	968
3	3	715
4	4	220
5	5	719
6	6	252
7	7	150
8	8	434

16 An advertising agency set up an experiment to see what features of an automobile influenced consumers to buy. As part of the experiment they asked a group of automotive engineers to rank 10 cars, and they asked a group of typical consumers to rank the same 10 cars. Test at $\alpha = .05$ to see if there is significant correlation between the two sets of rankings. (First convert points to ranks.)

Car	Points Awarded by Expert	Points Awarded by Consumer
1	9	3
2	5	4
3	1	3
4	5	3
5	7	6
6	6	5
7	2	9
8	5	9
9	5	8
10	2	5

17 A behavioral psychologist was studying the relation between overcrowding and violent behavior in a rat colony. She established several colonies with different degrees of crowding by putting more or fewer rats in a cage. She recorded what she considered to be acts of violence. She then ranked each colony for crowdedness and violence. A rank of 1 meant most crowded or most violent. A rank of 8 meant least crowded or least violent. Test at $\alpha = .05$ for significant correlation.

Colony	Population Density Rank	Violence Rank
1	1	1.5
2	2	1.5
3	3	4
4	4	3
5	5	5
6	6	7
7	7	7
8	8	7

18 A union representative was studying tipping patterns in large restaurants before entering into contract negotiations. She was checking to see if there was a correlation between a waiter's experience and how well he did in tips. At random, 14 waiters were picked from several comparable large restaurants in Manhattan and then ranked by experience and amount of tips per week. Here are the results. (Number 1 rank in experience represents the most experience. Number 1 rank in tips represents the most tips.) With $\alpha = .05$ is there significant correlation?

Waiter	Experience Rank	Tips Rank
Aitken	1	4
Anderson	2	2
Anscombe	3	6
Belayev	4	5
Birnbaum	5	1
Chow	6	3
Derman	6	7
Franklin	8	9
Harris	9	13
Kawata	10	14
Renyi	11	11
Sikorski	12	8
Teicher	13	10
Welch	14	12

19 A professor of education, Tab LaRasa, suspected that there was no correlation between the grades his students got on the final exam and the amount of time they spent studying for it during the week preceding the exam. He collected data for several semesters. Here is a random sample for 30 of his students. Does the data indicate at the .05 significance level that Professor LaRasa's intuition was correct?

Student	Number of Hours Studied for Exam	Grade on Exam	Student	Number of Hours Studied for Exam	Grade on Exam
1	21	95	16	20	70
2	18	72	17	18	86
3	1	61	18	2	75
4	9	73	19	12	75
5	13	68	20	20	59
6	1	52	21	15	66
7	17	60	22	11	67
8	2	86	23	21	84
9	19	67	24	14	78
10	1	64	25	13	81
11	6	73	26	1	75
12	19	76	27	3	86
13	14	53	28	6	81
14	4	69	29	2	59
15	0	90	30	2	58

a Compute r, the Pearson correlation coefficient, and test for nonzero correlation.

b Compute r_s, the rank correlation coefficient, and test for nonzero correlation.

20 To investigate correlation between cigarette smoking and absenteeism in a large factory, 30 employees were selected at random, and data was obtained on these two variables. Convert the data to rank measure and test for significant correlation. Use $\alpha = .05$.

Employee	Cigarettes per Day	Days Absent per Year	Employee	Cigarettes per Day	Days Absent per Year
1	11	6	16	9	9
2	5	4	17	9	21
3	8	4	18	7	9
4	39	3	19	5	1
5	9	8	20	8	0
6	0	6	21	8	8
7	3	4	22	25	18
8	7	6	23	6	6
9	2	5	24	5	1
10	5	0	25	27	2
11	0	7	26	5	5
12	32	6	27	3	7
13	7	9	28	7	2
14	5	1	29	9	5
15	20	7	30	15	13

21 Officials of the racing association were testing the effect of a drug on race horse performance. They chose 10 horses of similar ability, as judged by their past performance, and randomly split them into two groups of 5 horses. One group was given an injection of the drug; the other was given a neutral injection. The jockeys were not told which injection their horse had received. Then the 10 horses were raced on a straight track for $\frac{1}{2}$ mile. The results are given in the table. Test at $\alpha = .05$ to compare the two groups of horses. Is it likely that the drug affects performance? Would it have been a better idea to tell each jockey which injection his horse received?

Horse	Given Drug	Finish in Race
My Pal	Yes	2
Your Pal	No	10
Her Pal	No	7
His Pal	No	9
Go Fast	Yes	1
Win Big	Yes	4
Fleet	Yes	6
Baby Doll	Yes	3
Horse	No	8
My Hobby	No	5

22 a Devise a distribution-free experiment of your own.
b Carry out the experiment that you described in part a.

23 Make up your own question and use the data in Appendix A to answer it.

Readings from J. Tanur (ed.), "Statistics: A Guide to the Unknown," Holden-Day, San Francisco, 1972.

24 Preliminary Evaluation of a New Food Product, Street and Carroli, p. 220. Converting picture rankings to usable form.

25 The Meaning of Words, Kruskal, p. 185. What is multidimensional scaling? What is an agreement score?

Review

26 100 athletes were questioned as to their grades and their sport season. Of 50 athletes who participated in fall sports, 22 said that their grades were higher in the fall, 18 said that they were higher in the spring, and 10 said that there was no difference. Of 50 athletes who participated in spring sports, 21 said that their grades were higher in the fall, 17 said that they were higher in the spring, and 12 said that there was no difference. Do the data indicate that the season of play is related to the grade pattern? What is the hypothesis of independence? Perform a hypothesis test on this data at the .05 significance level.

27 The Orlando Star stated that 60 percent of community college students go on to 4-year schools. A student at Normal Community College interviewed 85 fellow students and 36 said that they intended to go on to a 4-year school. Is this sufficient evidence to say that the 60 percent figure does not apply at Normal Community College? Use $\alpha = .05$.

28 A student wanted to discover if there was any difference between the percentage of day students who favored euthanasia and the percentage of evening students who favored euthanasia. Her samples showed that 30 out of 58 day students and 24 out of 42 evening students favored euthanasia. Using $\alpha = .01$, does this suggest that the attitudes are different?

29 A survey taken in Monkey Ward's offices showed that 90 men employees smoked an average of 16.8 cigarettes between 9 A.M. and 5 P.M. The standard deviation was 11.3. At the same time 50 women employees averaged 20.6 cigarettes with a standard deviation of 13.2. Does this evidence indicate any difference between the number of cigarettes smoked between 9 A.M. and 5 P.M. by the two sexes? Let $\alpha = .05$.

30 a Estimate the average height of boys in the third grade if 10 classmates at a birthday party were the following heights (assume the population is normal):

Billy	53	Danny	56
Mark	53	Steven	54
Juan	48.25	Douglas	54
David	54	Marc	56
Thomas	49.5	Michael	57

b Find the 99 percent confidence interval for your estimate.

31 a Using the data below, estimate the difference between the percentage of males on campus who favor the legalization of marijuana and the percentage of females on campus who do.

b Find a 99 percent confidence interval for your estimate.

Data: 80 out of 100 males and 75 out of 100 females favored the legalization.

32 a A student wished to see if there was any correlation between marks on the first test in his economics class and marks on the first test in his English class. He sampled five students who were in both classes. Their marks were:

Student	Economics	English
Steven	76	60
Bob	82	72
Skip	80	68
Cheryl	90	84
Danny	84	66

Find the sample coefficient of correlation and test for correlation at the .05 significance level.

b Write the prediction formula.

c If Sue received an 81 in economics, what do you predict that she would get if she had been in the English class?

APPENDIX A:

Sample Data

A random sample of students at Nassau Community College was given the questionnaire below. The answers were gathered into the table that follows. It is given here as a source of sample data for both the teacher and the student. A variety of questions can be investigated using this data.

The data is given as it was actually collected. Thus the student must face the problem and decide what to do when a person does not answer one question (see 3), or refuses to answer any questions "on principle" (see 74), or even answers twice (see 26). What should one do about seemingly contradictory answers (see 36)? How might the questionnaire be improved?

QUESTIONNAIRE

Please check *one* response for each item.

I. Sex: ________ male ________ female

II. Marital Status: ________ single ________ married
________ widowed ________ divorced

III. Political Affiliation: ________ conservative
________ democrat ________ republican ________ liberal
________ other

IV. Religion: ________ Protestant ________ Catholic
________ Jewish ________ other ________ none

V. Have you attended any other college? ________ yes
________ no

VI. Did you vote in the last public election? ________ yes
________ no ________ ineligible

VII. Are you a full-time student? ________ yes ________ no

VIII. How many hours per week do you work? (a) ________ none
(b) ________ under 10 (c) ________ 10 to 20
(d) ________ over 20

IX. Are you a: ________ freshman ________ sophomore
________ junior ________ senior

X. How many extra-curricular school activities are you a member of:
(a) ________ none (b) ________ one (c) ________ two
(d) ________ more than two

Please fill in the appropriate response

XI. Please give your age correct to the *nearest* month:
________ years ________ months

XII. Please give your grade point average: ________

XIII. How much money did you earn last year, correct to the nearest thousand dollars? $______,000.

	I	II	III	IV	V	VI	VII	VIII	IX	X	XI	XII	XIII
1	F	M	D	P	Y	Y	Y	a	s	a	42, 8	4.0	0
2	F	S	D	J	N	Y	Y	d	s	a	18, 1	3.66	0
3	F	M	R	P	Y	Y	N	d	. . .	a	45, 4	. . .	12
4	M	S	D	N	N	Y	Y	a	S	a	35, 2	2.8	0
5	F	M	L	J	Y	Y	Y	d	j	a	39, 6	3.0	0
6	F	S	L	P	N	Y	Y	b	s	b	19, 10	2.30	0
7	M	S	R	C	N	N	Y	d	f	a	19, 5	2.8	1
8	F	M	C	C	N	Y	N	d	j	a	40, 0	3.8	0
9	M	S	O	N	Y	Y	Y	a	s	b	20, 6	3.29	0
10	M	M	O	C	N	N	Y	d	s	a	28, 4	. . .	13
11	F	M	D	J	Y	Y	Y	a	S	a	31, 0	3.7	0
12	M	S	D	C	N	Y	Y	d	S	a	20, 1	2.7	7
13	F	S	R	C	N	Y	Y	c	s	a	18, 11	2.57	1
14	M	S	D	N	N	Y	N	c	s	a	21, 6	2.5	4
15	F	S	L	C	N	Y	Y	c	s	a	19, 3	3.35	1
16	M	M	D	N	N	Y	N	d	s	a	44, 8	3.95	23
17	F	S	R	J	N	Y	Y	b	s	a	19, 3	3.0	1
18	M	S	D	P	N	Y	Y	a	f	a	21, 4	3.0	10
19	M	S	C	C	N	Y	Y	d	a	a	14, 7	. . .	0
20	M	S	R	C	Y	Y	Y	d	S	a	21, 3	3.0	4
21	M	S	D	C	N	Y	Y	b	j	b	25, 1	2.78	1
22	F	M	L	N	Y	Y	Y	c	j	a	26, 9	3.4	6
23	F	S	R	C	N	Y	Y	b	s	a	20, 10	2.5	1
24	M	S	L	C	Y	Y	Y	c	s	a	20, 0	2.84	4
25	F	S	O	C	N	Y	Y	d	s	b	19, 1	3.1	1
26	M	S	. . .	C,J	N	N	N	d	s	a	20, 8	2.0	3
27	F	S	R	C	Y	N	Y	d	j	b	20, 9	2.0	1
28	M	S	D	C	N	Y	Y	c	f	a	23, 2	. . .	0
29	F	S	D	C	Y	Y	Y	d	s	d	18, 11	3.8	2
30	M	S	O	N	Y	Y	Y	a	S	a	21, 8	3.1	1
31	F	M	D	J	N	Y	N	a	f	a	27, 1	. . .	0
32	F	S	. . .	C	N	Y	Y	d	f	. . .	19, 1	2.5	2
33	M	S	R	C	N	Y	Y	d	s	a	19, 6	. . .	0
34	M	M	L	C	Y	Y	Y	d	s	a	23, 9	2.8	11
35	F	S	R	J	N	I	Y	a	s	a	18, 9	2.5	0
36	M	M	O	N	Y	Y	N	a	. . .	a	26, 5	. . .	12
37	F	S	R	C	N	Y	Y	d	j	a	18, 11	. . .	0
38	M	S	L	C	Y	Y	N	d	s	b	19, 5	2.9	1
39	F	S	O	N	Y	Y	Y	b	j	d	20, 1	2.9	0
40	M	S	D	J	N	Y	Y	c	s	b	18, 11	. . .	1
41	F	S	L	C	Y	Y	Y	d	j	b	19, 10	3.2	1
42	M	S	L	C	N	Y	Y	d	f	a	18, 6	. . .	3
43	M	S	O	N	Y	Y	Y	c	s	c	19, 5	2.5	1
44	M	S	R	J	N	I	Y	b	f	d	17, 2	. . .	0
45	M	S	L	J	Y	N	Y	b	s	b	23	3.0	4
46	M	S	D	P	Y	Y	Y	c	s	a	21, 0	2.5	1
47	F	S	O	O	N	N	Y	c	s	a	20, 3	3.0	0
48	M	M	D	C	N	Y	Y	b	j	d	26, 8	3.75	3
49	F	S	D	J	Y	Y	Y	d	s	c	19, 1	2.8	1
50	M	S	R	C	N	I	Y	a	f	c	18, 3	. . .	1
51	F	S	L	J	N	N	N	a	f	a	17, 0	. . .	0
52	M	S	O	C	Y	Y	Y	d	s	a	19, 9	3.5	2
53	M	S	O	C	N	I	Y	c	f	a	18, 9	. . .	0
54	M	S	R	C	Y	Y	Y	b	s	b	21, 4	2.7	1
55	F	S	R	C	Y	Y	Y	c	s	c	18, 11	3.5	0

s = sophomore; S = senior

(Continued)

	I	II	III	IV	V	VI	VII	VIII	IX	X	XI	XII	XIII
56	M	S	O	C	N	N	N	d	f	a	26, 9	. . .	19
57	M	S	D	N	Y	Y	Y	d	s	b	19, 11	3.19	4
58	F	S	. . .	J	N	I	N	c	f	c	17, 9	. . .	0
59	F	S	O	O	Y	I	Y	b	s	a	18, 9	3.5	1
60	F	S	R	C	N	Y	Y	c	s	a	19, 0	3.0	1
61	M	S	L	C	N	Y	Y	c	s	a	18, 11	2.46	0
62	F	S	D	J	Y	Y	Y	c	s	a	19, 2	. . .	1
63	F	S	D	C	N	Y	Y	d	s	a	20, 0	2.4	5
64	M	M	. . .	C	Y	N	Y	a	s	a	26, 2	3.19	1
54	M	M	C	C	N	N	Y	c	s	a	25, 3	3.7	10
66	M	S	O	P	N	I	Y	b	s	a	18, 9	2.96	0
67	M	S	O	J	Y	Y	N	d	. . .	c	24, 3	3.56	10
68	F	S	O	J	Y	Y	Y	b	S	d	21, 5	3.45	1
69	F	S	D	J	N	I	Y	d	f	a	18, 6	. . .	1
70	M	S	D	C	N	I	Y	a	f	a	17, 8	. . .	0
71	M	S	O	J	Y	Y	Y	c	S	b	22, 1	2.35	1
72	F	S	D	J	N	I	Y	c	s	a	18, 1	3.73	0
73	F	S	D	C	Y	Y	Y	a	S	a	22, 2	3.8	0
74	. . .	. . .	. . .	. . .	. . .	. . .	. . .	. . .	. . .	. . .	. . .	. . .	. . .

ASSIGNMENTS BASED ON THE QUESTIONNAIRE

I Chi square

1 Are sex and religion independent in this population?

2 Is there a relationship between full-time status and number of hours worked?

II Binomial

1 Is the percentage of married Catholic students the same as the percentage of married Jewish students?

2 Find the 95 percent confidence interval for the percentage of full-time students who are married.

III Sample mean

1 Estimate the average age of married students. Find the 99 percent confidence interval for your estimate.

2 Estimate the difference between the means of the grade point averages of single and married students.

APPENDIX B:

Tables

B-1 Chi square
B-2 Square roots
B-3a A short table of areas under a normal curve to the left of z
B-3b A more complete table of areas under a normal curve to the left of z
B-4 Critical values of t for a two-tailed test
B-5 Critical values of t for a one-tailed test
B-6 Coefficients of correlation r_c for a two-tailed test
B-7 Coefficients of correlation r_c for a one-tailed test
B-8 Critical values for rank sum
B-9 Critical values for rank correlation coefficient r_s

TABLE B-1 CHI SQUARE

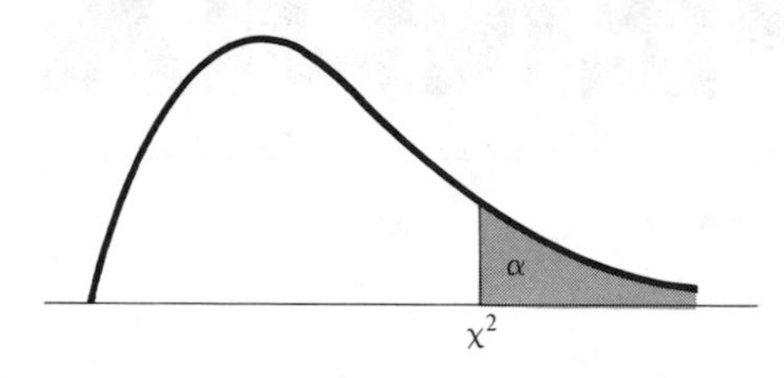

Degrees of Freedom	X_c^2 for $\alpha = .05$	X_c^2 for $\alpha = .01$	Degrees of Freedom	X_c^2 for $\alpha = .05$	X_c^2 for $\alpha = .01$
1	3.84	6.635	16	26.30	32.00
2	5.99	9.21	17	27.59	33.41
3	7.82	11.34	18	28.87	34.80
4	9.49	13.28	19	30.14	36.19
5	11.07	15.09	20	31.41	37.57
6	12.59	16.81	21	32.67	38.93
7	14.07	18.48	22	33.92	40.29
8	15.51	20.09	23	35.17	41.64
9	16.92	21.67	24	36.42	42.98
10	18.31	23.21	25	37.65	44.31
11	19.68	24.72	26	38.88	45.64
12	21.03	26.22	27	40.11	46.96
13	22.36	27.69	28	41.34	48.28
14	23.68	29.14	29	42.56	49.59
15	25.00	30.58	30	43.77	50.89

Note: For critical values of X^2 when the degrees of freedom exceed 30, we can approximate X_c^2 by the formula

$$X_c^2 = \frac{1}{2}(z_c + \sqrt{2d - 1})^2$$

where d is the degrees of freedom and z_c is either 1.65 for $\alpha = .05$ or 2.33 for $\alpha = .01$.

TABLE B-2 SQUARE ROOTS

N	$\sqrt{N}$	$\sqrt{10N}$	N	$\sqrt{N}$	$\sqrt{10N}$	N	$\sqrt{N}$	$\sqrt{10N}$
1.00	1.000	3.162	1.50	1.225	3.873	2.00	1.414	4.472
1.01	1.005	3.178	1.51	1.229	3.886	2.01	1.418	4.483
1.02	1.010	3.194	1.52	1.233	3.899	2.02	1.421	4.494
1.03	1.015	3.209	1.53	1.237	3.912	2.03	1.425	4.506
1.04	1.020	3.225	1.54	1.241	3.924	2.04	1.428	4.517
1.05	1.025	3.240	1.55	1.245	3.937	2.05	1.432	4.528
1.06	1.030	3.256	1.56	1.249	3.950	2.06	1.435	4.539
1.07	1.034	3.271	1.57	1.253	3.962	2.07	1.439	4.550
1.08	1.039	3.286	1.58	1.257	3.975	2.08	1.442	4.561
1.09	1.044	3.302	1.59	1.261	3.987	2.09	1.446	4.572
1.10	1.049	3.317	1.60	1.265	4.000	2.10	1.449	4.583
1.11	1.054	3.332	1.61	1.269	4.012	2.11	1.453	4.593
1.12	1.058	3.347	1.62	1.273	4.025	2.12	1.456	4.604
1.13	1.063	3.362	1.63	1.277	4.037	2.13	1.459	4.615
1.14	1.068	3.376	1.64	1.281	4.050	2.14	1.463	4.626
1.15	1.072	3.391	1.65	1.285	4.062	2.15	1.466	4.637
1.16	1.077	3.406	1.66	1.288	4.074	2.16	1.470	4.648
1.17	1.082	3.421	1.67	1.292	4.087	2.17	1.473	4.658
1.18	1.086	3.435	1.68	1.296	4.099	2.18	1.476	4.669
1.19	1.091	3.450	1.69	1.300	4.111	2.19	1.480	4.680
1.20	1.095	3.464	1.70	1.304	4.123	2.20	1.483	4.690
1.21	1.100	3.479	1.71	1.308	4.135	2.21	1.487	4.701
1.22	1.105	3.493	1.72	1.311	4.147	2.22	1.490	4.712
1.23	1.109	3.507	1.73	1.315	4.159	2.23	1.493	4.722
1.24	1.114	3.521	1.74	1.319	4.171	2.24	1.497	4.733
1.25	1.118	3.536	1.75	1.323	4.183	2.25	1.500	4.743
1.26	1.122	3.550	1.76	1.327	4.195	2.26	1.503	4.754
1.27	1.127	3.564	1.77	1.330	4.207	2.27	1.507	4.764
1.28	1.131	3.578	1.78	1.334	4.219	2.28	1.510	4.775
1.29	1.136	3.592	1.79	1.338	4.231	2.29	1.513	4.785
1.30	1.140	3.606	1.80	1.342	4.243	2.30	1.517	4.796
1.31	1.145	3.619	1.81	1.345	4.254	2.31	1.520	4.806
1.32	1.149	3.633	1.82	1.349	4.266	2.32	1.523	4.817
1.33	1.153	3.647	1.83	1.353	4.278	2.33	1.526	4.827
1.34	1.158	3.661	1.84	1.356	4.290	2.34	1.530	4.837
1.35	1.162	3.674	1.85	1.360	4.301	2.35	1.533	4.848
1.36	1.166	3.688	1.86	1.364	4.313	2.36	1.536	4.858
1.37	1.170	3.701	1.87	1.367	4.324	2.37	1.539	4.868
1.38	1.175	3.715	1.88	1.371	4.336	2.38	1.543	4.879
1.39	1.179	3.728	1.89	1.375	4.347	2.39	1.546	4.889
1.40	1.183	3.742	1.90	1.378	4.359	2.40	1.549	4.899
1.41	1.187	3.755	1.91	1.382	4.370	2.41	1.552	4.909
1.42	1.192	3.768	1.92	1.386	4.382	2.42	1.556	4.919
1.43	1.196	3.782	1.93	1.389	4.393	2.43	1.559	4.930
1.44	1.200	3.795	1.94	1.393	4.405	2.44	1.562	4.940
1.45	1.204	3.808	1.95	1.396	4.416	2.45	1.565	4.950
1.46	1.208	3.821	1.96	1.400	4.427	2.46	1.568	4.960
1.47	1.212	3.834	1.97	1.404	4.438	2.47	1.572	4.970
1.48	1.217	3.847	1.98	1.407	4.450	2.48	1.575	4.980
1.49	1.221	3.860	1.99	1.411	4.461	2.49	1.578	4.990

TABLE B-2 (*Continued*)

N	$\sqrt{N}$	$\sqrt{10N}$	N	$\sqrt{N}$	$\sqrt{10N}$	N	$\sqrt{N}$	$\sqrt{10N}$
2.50	1.581	5.000	3.00	1.732	5.477	3.50	1.871	5.916
2.51	1.584	5.010	3.01	1.735	5.486	3.51	1.873	5.925
2.52	1.587	5.020	3.02	1.738	5.495	3.52	1.876	5.933
2.53	1.591	5.030	3.03	1.741	5.505	3.53	1.879	5.941
2.54	1.594	5.040	3.04	1.744	5.514	3.54	1.881	5.950
2.55	1.597	5.050	3.05	1.746	5.523	3.55	1.884	5.958
2.56	1.600	5.060	3.06	1.749	5.532	3.56	1.887	5.967
2.57	1.603	5.070	3.07	1.752	5.541	3.57	1.889	5.975
2.58	1.606	5.079	3.08	1.755	5.550	3.58	1.892	5.983
2.59	1.609	5.089	3.09	1.758	5.559	3.59	1.895	5.992
2.60	1.612	5.099	3.10	1.761	5.568	3.60	1.897	6.000
2.61	1.616	5.109	3.11	1.764	5.577	3.61	1.900	6.008
2.62	1.619	5.119	3.12	1.766	5.586	3.62	1.903	6.017
2.63	1.622	5.128	3.13	1.769	5.595	3.63	1.905	6.025
2.64	1.625	5.138	3.14	1.772	5.604	3.64	1.908	6.033
2.65	1.628	5.148	3.15	1.775	5.612	3.65	1.910	6.042
2.66	1.631	5.158	3.16	1.778	5.621	3.66	1.913	6.050
2.67	1.634	5.167	3.17	1.780	5.630	3.67	1.916	6.058
2.68	1.637	5.177	3.18	1.783	5.639	3.68	1.918	6.066
2.69	1.640	5.187	3.19	1.786	5.648	3.69	1.921	6.075
2.70	1.643	5.196	3.20	1.789	5.657	3.70	1.924	6.083
2.71	1.646	5.206	3.21	1.792	5.666	3.71	1.926	6.091
2.72	1.649	5.215	3.22	1.794	5.675	3.72	1.929	6.099
2.73	1.652	5.225	3.23	1.797	5.683	3.73	1.931	6.107
2.74	1.655	5.234	3.24	1.800	5.692	3.74	1.934	6.116
2.75	1.658	5.244	3.25	1.803	5.701	3.75	1.936	6.124
2.76	1.661	5.254	3.26	1.806	5.710	3.76	1.939	6.132
2.77	1.664	5.263	3.27	1.808	5.718	3.77	1.942	6.140
2.78	1.667	5.273	3.28	1.811	5.727	3.78	1.944	6.148
2.79	1.670	5.282	3.29	1.814	5.736	3.79	1.947	6.156
2.80	1.673	5.292	3.30	1.817	5.745	3.80	1.949	6.164
2.81	1.676	5.301	3.31	1.819	5.753	3.81	1.952	6.173
2.82	1.679	5.310	3.32	1.822	5.762	3.82	1.954	6.181
2.83	1.682	5.320	3.33	1.825	5.771	3.83	1.957	6.189
2.84	1.685	5.329	3.34	1.828	5.779	3.84	1.960	6.197
2.85	1.688	5.339	3.35	1.830	5.788	3.85	1.962	6.205
2.86	1.691	5.348	3.36	1.833	5.797	3.86	1.965	6.213
2.87	1.694	5.357	3.37	1.836	5.805	3.87	1.967	6.221
2.88	1.697	5.367	3.38	1.838	5.814	3.88	1.970	6.229
2.89	1.700	5.376	3.39	1.841	5.822	3.89	1.972	6.237
2.90	1.703	5.385	3.40	1.844	5.831	3.90	1.975	6.245
2.91	1.706	5.394	3.41	1.847	5.840	3.91	1.977	6.253
2.92	1.709	5.404	3.42	1.849	5.848	3.92	1.980	6.261
2.93	1.712	5.413	3.43	1.852	5.857	3.93	1.982	6.269
2.94	1.715	5.422	3.44	1.855	5.865	3.94	1.985	6.277
2.95	1.718	5.431	3.45	1.857	5.874	3.95	1.987	6.285
2.96	1.720	5.441	3.46	1.860	5.882	3.96	1.990	6.293
2.97	1.723	5.450	3.47	1.863	5.891	3.97	1.992	6.301
2.98	1.726	5.459	3.48	1.865	5.899	3.98	1.995	6.309
2.99	1.729	5.468	3.49	1.868	5.908	3.99	1.997	6.317

TABLE B-2 *(Continued)*

N	$\sqrt{N}$	$\sqrt{10N}$	N	$\sqrt{N}$	$\sqrt{10N}$	N	$\sqrt{N}$	$\sqrt{10N}$
4.00	2.000	6.325	4.50	2.121	6.708	5.00	2.236	7.071
4.01	2.002	6.332	4.51	2.124	6.716	5.01	2.238	7.078
4.02	2.005	6.340	4.52	2.126	6.723	5.02	2.241	7.085
4.03	2.007	6.348	4.53	2.128	6.731	5.03	2.243	7.092
4.04	2.010	6.356	4.54	2.131	6.738	5.04	2.245	7.099
4.05	2.012	6.364	4.55	2.133	6.745	5.05	2.247	7.106
4.06	2.015	6.372	4.56	2.135	6.753	5.06	2.249	7.113
4.07	2.017	6.380	4.57	2.138	6.760	5.07	2.252	7.120
4.08	2.020	6.387	4.58	2.140	6.768	5.08	2.254	7.127
4.09	2.022	6.395	4.59	2.142	6.775	5.09	2.256	7.134
4.10	2.025	6.403	4.60	2.145	6.782	5.10	2.258	7.141
4.11	2.027	6.411	4.61	2.147	6.790	5.11	2.261	7.148
4.12	2.030	6.419	4.62	2.149	6.797	5.12	2.263	7.155
4.13	2.032	6.427	4.63	2.152	6.804	5.13	2.265	7.162
4.14	2.035	6.434	4.64	2.154	6.812	5.14	2.267	7.169
4.15	2.037	6.442	4.65	2.156	6.819	5.15	2.269	7.176
4.16	2.040	6.450	4.66	2.159	6.826	5.16	2.272	7.183
4.17	2.042	6.458	4.67	2.161	6.834	5.17	2.274	7.190
4.18	2.045	6.465	4.68	2.163	6.841	5.18	2.276	7.197
4.19	2.047	6.473	4.69	2.166	6.848	5.19	2.278	7.204
4.20	2.049	6.481	4.70	2.168	6.856	5.20	2.280	7.211
4.21	2.052	6.488	4.71	2.170	6.863	5.21	2.283	7.218
4.22	2.054	6.496	4.72	2.173	6.870	5.22	2.285	7.225
4.23	2.057	6.504	4.73	2.175	6.877	5.23	2.287	7.232
4.24	2.059	6.512	4.74	2.177	6.885	5.24	2.289	7.239
4.25	2.062	6.519	4.75	2.179	6.892	5.25	2.291	7.246
4.26	2.064	6.527	4.76	2.182	6.899	5.26	2.293	7.253
4.27	2.066	6.535	4.77	2.184	6.907	5.27	2.296	7.259
4.28	2.069	6.542	4.78	2.186	6.914	5.28	2.298	7.266
4.29	2.071	6.550	4.79	2.189	6.921	5.29	2.300	7.273
4.30	2.074	6.557	4.80	2.191	6.928	5.30	2.302	7.280
4.31	2.076	6.565	4.81	2.193	6.935	5.31	2.304	7.287
4.32	2.078	6.573	4.82	2.195	6.943	5.32	2.307	7.294
4.33	2.081	6.580	4.83	2.198	6.950	5.33	2.309	7.301
4.34	2.083	6.588	4.84	2.200	6.957	5.34	2.311	7.308
4.35	2.086	6.595	4.85	2.202	6.964	5.35	2.313	7.314
4.36	2.088	6.603	4.86	2.205	6.971	5.36	2.315	7.321
4.37	2.090	6.611	4.87	2.207	6.979	5.37	2.317	7.328
4.38	2.093	6.618	4.88	2.209	6.986	5.38	2.319	7.335
4.39	2.095	6.626	4.89	2.211	6.993	5.39	2.322	7.342
4.40	2.098	6.633	4.90	2.214	7.000	5.40	2.324	7.348
4.41	2.100	6.641	4.91	2.216	7.007	5.41	2.326	7.355
4.42	2.102	6.648	4.92	2.218	7.014	5.42	2.328	7.362
4.43	2.105	6.656	4.93	2.220	7.021	5.43	2.330	7.369
4.44	2.107	6.663	4.94	2.223	7.029	5.44	2.332	7.376
4.45	2.110	6.671	4.95	2.225	7.036	5.45	2.335	7.382
4.46	2.112	6.678	4.96	2.227	7.043	5.46	2.337	7.389
4.47	2.114	6.686	4.97	2.229	7.050	5.47	2.339	7.396
4.48	2.117	6.693	4.98	2.232	7.057	5.48	2.341	7.403
4.49	2.119	6.701	4.99	2.234	7.064	5.49	2.343	7.409

TABLE B-2 (*Continued*)

N	$\sqrt{N}$	$\sqrt{10N}$	N	$\sqrt{N}$	$\sqrt{10N}$	N	$\sqrt{N}$	$\sqrt{10N}$
5.50	2.345	7.416	6.00	2.449	7.746	6.50	2.550	8.062
5.51	2.347	7.423	6.01	2.452	7.752	6.51	2.551	8.068
5.52	2.349	7.430	6.02	2.454	7.759	6.52	2.553	8.075
5.53	2.352	7.436	6.03	2.456	7.765	6.53	2.555	8.081
5.54	2.354	7.443	6.04	2.458	7.772	6.54	2.557	8.087
5.55	2.356	7.450	6.05	2.460	7.778	6.55	2.559	8.093
5.56	2.358	7.457	6.06	2.462	7.785	6.56	2.561	8.099
5.57	2.360	7.463	6.07	2.464	7.791	6.57	2.563	8.106
5.58	2.362	7.470	6.08	2.466	7.797	6.58	2.565	8.112
5.59	2.364	7.477	6.09	2.468	7.804	6.59	2.567	8.118
5.60	2.366	7.483	6.10	2.470	7.810	6.60	2.569	8.124
5.61	2.369	7.490	6.11	2.472	7.817	6.61	2.571	8.130
5.62	2.371	7.497	6.12	2.474	7.823	6.62	2.573	8.136
5.63	2.373	7.503	6.13	2.476	7.829	6.63	2.575	8.142
5.64	2.375	7.510	6.14	2.478	7.836	6.64	2.577	8.149
5.65	2.377	7.517	6.15	2.480	7.842	6.65	2.579	8.155
5.66	2.379	7.523	6.16	2.482	7.849	6.66	2.581	8.161
5.67	2.381	7.530	6.17	2.484	7.855	6.67	2.583	8.167
5.68	2.383	7.537	6.18	2.486	7.861	6.68	2.585	8.173
5.69	2.385	7.543	6.19	2.488	7.868	6.69	2.587	8.179
5.70	2.387	7.550	6.20	2.490	7.874	6.70	2.588	8.185
5.71	2.390	7.556	6.21	2.492	7.880	6.71	2.590	8.191
5.72	2.392	7.563	6.22	2.494	7.887	6.72	2.592	8.198
5.73	2.394	7.570	6.23	2.496	7.893	6.73	2.594	8.204
5.74	2.396	7.576	6.24	2.498	7.899	6.74	2.596	8.210
5.75	2.398	7.583	6.25	2.500	7.906	6.75	2.598	8.216
5.76	2.400	7.589	6.26	2.502	7.912	6.76	2.600	8.222
5.77	2.402	7.596	6.27	2.504	7.918	6.77	2.602	8.228
5.78	2.404	7.603	6.28	2.506	7.925	6.78	2.604	8.234
5.79	2.406	7.609	6.29	2.508	7.931	6.79	2.606	8.240
5.80	2.408	7.616	6.30	2.510	7.937	6.80	2.608	8.246
5.81	2.410	7.622	6.31	2.512	7.944	6.81	2.610	8.252
5.82	2.412	7.629	6.32	2.514	7.950	6.82	2.612	8.258
5.83	2.415	7.635	6.33	2.516	7.956	6.83	2.613	8.264
5.84	2.417	7.642	6.34	2.518	7.962	6.84	2.615	8.270
5.85	2.419	7.649	6.35	2.520	7.969	6.85	2.617	8.276
5.86	2.421	7.655	6.36	2.522	7.975	6.86	2.619	8.283
5.87	2.423	7.662	6.37	2.524	7.981	6.87	2.621	8.289
5.88	2.425	7.668	6.38	2.526	7.987	6.88	2.623	8.295
5.89	2.427	7.675	6.39	2.528	7.994	6.89	2.625	8.301
5.90	2.429	7.681	6.40	2.530	8.000	6.90	2.627	8.307
5.91	2.431	7.688	6.41	2.532	8.006	6.91	2.629	8.313
5.92	2.433	7.694	6.42	2.534	8.012	6.92	2.631	8.319
5.93	2.435	7.701	6.43	2.536	8.019	6.93	2.632	8.325
5.94	2.437	7.707	6.44	2.538	8.025	6.94	2.634	8.331
5.95	2.439	7.714	6.45	2.540	8.031	6.95	2.636	8.337
5.96	2.441	7.720	6.46	2.542	8.037	6.96	2.638	8.343
5.97	2.443	7.727	6.47	2.544	8.044	6.97	2.640	8.349
5.98	2.445	7.733	6.48	2.546	8.050	6.98	2.642	8.355
5.99	2.447	7.740	6.49	2.548	8.056	6.99	2.644	8.361

TABLE B-2 *(Continued)*

N	$\sqrt{N}$	$\sqrt{10N}$	N	$\sqrt{N}$	$\sqrt{10N}$	N	$\sqrt{N}$	$\sqrt{10N}$
7.00	2.646	8.367	7.50	2.739	8.660	8.00	2.828	8.944
7.01	2.648	8.373	7.51	2.740	8.666	8.01	2.830	8.950
7.02	2.650	8.379	7.52	2.742	8.672	8.02	2.832	8.955
7.03	2.651	8.385	7.53	2.744	8.678	8.03	2.834	8.961
7.04	2.653	8.390	7.54	2.746	8.683	8.04	2.835	8.967
7.05	2.655	8.396	7.55	2.748	8.689	8.05	2.837	8.972
7.06	2.657	8.402	7.56	2.750	8.695	8.06	2.839	8.978
7.07	2.659	8.408	7.57	2.751	8.701	8.07	2.841	8.983
7.08	2.661	8.414	7.58	2.753	8.706	8.08	2.843	8.989
7.09	2.663	8.420	7.59	2.755	8.712	8.09	2.844	8.994
7.10	2.665	8.426	7.60	2.757	8.718	8.10	2.846	9.000
7.11	2.666	8.432	7.61	2.759	8.724	8.11	2.848	9.006
7.12	2.668	8.438	7.62	2.760	8.729	8.12	2.850	9.011
7.13	2.670	8.444	7.63	2.762	8.735	8.13	2.851	9.017
7.14	2.672	8.450	7.64	2.764	8.741	8.14	2.853	9.022
7.15	2.674	8.456	7.65	2.766	8.746	8.15	2.855	9.028
7.16	2.676	8.462	7.66	2.768	8.752	8.16	2.857	9.033
7.17	2.678	8.468	7.67	2.769	8.758	8.17	2.858	9.039
7.18	2.680	8.473	7.68	2.771	8.764	8.18	2.860	9.044
7.19	2.681	8.479	7.69	2.773	8.769	8.19	2.862	9.050
7.20	2.683	8.485	7.70	2.775	8.775	8.20	2.864	9.055
7.21	2.685	8.491	7.71	2.777	8.781	8.21	2.865	9.061
7.22	2.687	8.497	7.72	2.778	8.786	8.22	2.867	9.066
7.23	2.689	8.503	7.73	2.780	8.792	8.23	2.869	9.072
7.24	2.691	8.509	7.74	2.782	8.798	8.24	2.871	9.077
7.25	2.693	8.515	7.75	2.784	8.803	8.25	2.872	9.083
7.26	2.694	8.521	7.76	2.786	8.809	8.26	2.874	9.088
7.27	2.696	8.526	7.77	2.787	8.815	8.27	2.876	9.094
7.28	2.698	8.532	7.78	2.789	8.820	8.28	2.877	9.099
7.29	2.700	8.538	7.79	2.791	8.826	8.29	2.879	9.105
7.30	2.702	8.544	7.80	2.793	8.832	8.30	2.881	9.110
7.31	2.704	8.550	7.81	2.795	8.837	8.31	2.883	9.116
7.32	2.706	8.556	7.82	2.796	8.843	8.32	2.884	9.121
7.33	2.707	8.562	7.83	2.798	8.849	8.33	2.886	9.127
7.34	2.709	8.567	7.84	2.800	8.854	8.34	2.888	9.132
7.35	2.711	8.573	7.85	2.802	8.860	8.35	2.890	9.138
7.36	2.713	8.579	7.86	2.804	8.866	8.36	2.891	9.143
7.37	2.715	8.585	7.87	2.805	8.871	8.37	2.893	9.149
7.38	2.717	8.591	7.88	2.807	8.877	8.38	2.895	9.154
7.39	2.718	8.597	7.89	2.809	8.883	8.39	2.897	9.160
7.40	2.720	8.602	7.90	2.811	8.888	8.40	2.898	9.165
7.41	2.722	8.608	7.91	2.812	8.894	8.41	2.900	9.171
7.42	2.724	8.614	7.92	2.814	8.899	8.42	2.902	9.176
7.43	2.726	8.620	7.93	2.816	8.905	8.43	2.903	9.182
7.44	2.728	8.626	7.94	2.818	8.911	8.44	2.905	9.187
7.45	2.729	8.631	7.95	2.820	8.916	8.45	2.907	9.192
7.46	2.731	8.637	7.96	2.821	8.922	8.46	2.909	9.198
7.47	2.733	8.643	7.97	2.823	8.927	8.47	2.910	9.203
7.48	2.735	8.649	7.98	2.825	8.933	8.48	2.912	9.209
7.49	2.737	8.654	7.99	2.827	8.939	8.49	2.914	9.214

TABLE B-2 (*Continued*)

N	$\sqrt{N}$	$\sqrt{10N}$	N	$\sqrt{N}$	$\sqrt{10N}$	N	$\sqrt{N}$	$\sqrt{10N}$
8.50	2.915	9.220	9.00	3.000	9.487	9.50	3.082	9.747
8.51	2.917	9.225	9.01	3.002	9.492	9.51	3.084	9.752
8.52	2.919	9.230	9.02	3.003	9.497	9.52	3.085	9.757
8.53	2.921	9.236	9.03	3.005	9.503	9.53	3.087	9.762
8.54	2.922	9.241	9.04	3.007	9.508	9.54	3.089	9.767
8.55	2.924	9.247	9.05	3.008	9.513	9.55	3.090	9.772
8.56	2.926	9.252	9.06	3.010	9.518	9.56	3.092	9.778
8.57	2.927	9.257	9.07	3.012	9.524	9.57	3.094	9.783
8.58	2.929	9.263	9.08	3.013	9.529	9.58	3.095	9.788
8.59	2.931	9.268	9.09	3.015	9.534	9.59	3.097	9.793
8.60	2.933	9.274	9.10	3.017	9.539	9.60	3.098	9.798
8.61	2.934	9.279	9.11	3.018	9.545	9.61	3.100	9.803
8.62	2.936	9.284	9.12	3.020	9.550	9.62	3.102	9.808
8.63	2.938	9.290	9.13	3.022	9.555	9.63	3.103	9.813
8.64	2.939	9.295	9.14	3.023	9.560	9.64	3.105	9.818
8.65	2.941	9.301	9.15	3.025	9.566	9.65	3.106	9.823
8.66	2.943	9.306	9.16	3.027	9.571	9.66	3.108	9.829
8.67	2.944	9.311	9.17	3.028	9.576	9.67	3.110	9.834
8.68	2.946	9.317	9.18	3.030	9.581	9.68	3.111	9.839
8.69	2.948	9.322	9.19	3.031	9.586	9.69	3.113	9.844
8.70	2.950	9.327	9.20	3.033	9.592	9.70	3.114	9.849
8.71	2.951	9.333	9.21	3.035	9.597	9.71	3.116	9.854
8.72	2.953	9.338	9.22	3.036	9.602	9.72	3.118	9.859
8.73	2.955	9.343	9.23	3.038	9.607	9.73	3.119	9.864
8.74	2.956	9.349	9.24	3.040	9.612	9.74	3.121	9.869
8.75	2.958	9.354	9.25	3.041	9.618	9.75	3.122	9.874
8.76	2.960	9.359	9.26	3.043	9.623	9.76	3.124	9.879
8.77	2.961	9.365	9.27	3.045	9.628	9.77	3.126	9.884
8.78	2.963	9.370	9.28	3.046	9.633	9.78	3.127	9.889
8.79	2.965	9.375	9.29	3.048	9.638	9.79	3.129	9.894
8.80	2.966	9.381	9.30	3.050	9.644	9.80	3.130	9.899
8.81	2.968	9.386	9.31	3.051	9.649	9.81	3.132	9.905
8.82	2.970	9.391	9.32	3.053	9.654	9.82	3.134	9.910
8.83	2.972	9.397	9.33	3.055	9.659	9.83	3.135	9.915
8.84	2.973	9.402	9.34	3.056	9.664	9.84	3.137	9.920
8.85	2.975	9.407	9.35	3.058	9.670	9.85	3.138	9.925
8.86	2.977	9.413	9.36	3.059	9.675	9.86	3.140	9.930
8.87	2.978	9.418	9.37	3.061	9.680	9.87	3.142	9.935
8.88	2.980	9.423	9.38	3.063	9.685	9.88	3.143	9.940
8.89	2.982	9.429	9.39	3.064	9.690	9.89	3.145	9.945
8.90	2.983	9.434	9.40	3.066	9.695	9.90	3.146	9.950
8.91	2.985	9.439	9.41	3.068	9.701	9.91	3.148	9.955
8.92	2.987	9.445	9.42	3.069	9.706	9.92	3.150	9.960
8.93	2.988	9.450	9.43	3.071	9.711	9.93	3.151	9.965
8.94	2.990	9.455	9.44	3.072	9.716	9.94	3.153	9.970
8.95	2.992	9.460	9.45	3.074	9.721	9.95	3.154	9.975
8.96	2.993	9.466	9.46	3.076	9.726	9.96	3.156	9.980
8.97	2.995	9.471	9.47	3.077	9.731	9.97	3.158	9.985
8.98	2.997	9.476	9.48	3.079	9.737	9.98	3.159	9.990
8.99	2.998	9.482	9.49	3.081	9.742	9.99	3.161	9.995

HOW TO USE THE SQUARE ROOT TABLE

This table gives the square root of the numbers from 1.00 to 9.99 and also of the numbers from 10.0 to 99.9.

1 To find the square root of any number between 1.00 and 9.99 simply find that number in the column labeled N and read the square root in the column marked $\sqrt{N}$. For example, $\sqrt{2.24} = 1.497$.

2 To find the square root of a number between 10.0 and 99.9, such as 22.4, ignore the decimal point and locate the digits 224 in the column labeled N. Now read the square root from the column marked $\sqrt{10N}$. Thus we find $\sqrt{22.4} = 4.733$.

3 In general, we can find the square root of *any number* by using the following procedure.

i Make a *rough* estimate of the size of the answer. Is it near 10, 300, .04? What we need here is only the value of the first nonzero digit and the location of the decimal point. A good rule of thumb is to underline *two* numbers at a time starting at the decimal point (adding a zero if necessary). Thus,

a 360 becomes $\underline{03}\ \ \underline{60}$
b 4120 becomes $\underline{41}\ \ \underline{20}$
c .006 becomes $.\underline{00}\ \ \underline{60}$
d .000006 becomes $.\underline{00}\ \ \underline{00}\ \ \underline{06}$

Now estimate the first digit of the answer by finding the square root of the first nonzero underlined pair of digits. Thus

a $\sqrt{03}$ is approximately 1
b $\sqrt{41}$ is approximately 6
c $\sqrt{60}$ is approximately 7
d $\sqrt{06}$ is approximately 2

Finally, each *two* digits underlined will correspond to *one* digit in the answer. Thus we have

a $\sqrt{\underline{03}\ \ \underline{06}}$ is about 1________.; and so $\sqrt{360}$ is a number from 10 to 19
b $\sqrt{\underline{41}\ \ \underline{20}}$ is about 6________.; so $\sqrt{4,120}$ is a number in the sixties.
c $\sqrt{.\underline{00}\ \ \underline{60}}$ is about .07.
d $\sqrt{.\underline{00}\ \ \underline{00}\ \ \underline{06}}$ is about .002.

ii Now that you have estimated size of the answer, look up the square root of the three-digit number formed by ignoring the decimal point, ignoring any leading zeros, and adding zeros at the end if necessary. Thus we look up

a $\sqrt{360}$
b $\sqrt{412}$
c $\sqrt{600}$
d $\sqrt{600}$

Next to each of these numbers are *two* answers, but only one of them *begins* with the digit that we have already estimated. Select this answer and move the decimal point to correspond to your estimate. Thus

a $\sqrt{\underline{03}\ \ \underline{60}} = 1$________. $= 18.97$
b $\sqrt{\underline{41}\ \ \underline{20}} = 6$________. $= 64.19$

c $\sqrt{.\underline{00}\ \ \underline{60}} = .07\_\_\_\_ = .07746$
d $\sqrt{\underline{00}\ \ \underline{00}\ \ \underline{06}} = .002\_\_\_\_ = .002449$

Note 1: If a number has more than three significant digits, you can find its approximate square root in this table if you first round it off to three significant digits. Thus

$$\sqrt{8{,}428} = \sqrt{8{,}430} = \sqrt{\underline{84}\ \ \underline{30}} = 9\_\_\_\_. = 91.82$$

Similarly,

$$\sqrt{.01659} = \sqrt{.0166} = \sqrt{.\underline{01}\ \ \underline{66}} = .1\_\_\_\_. = .1288$$

Note 2: If time permits, it is a good idea to check your answer by squaring it. $\sqrt{8{,}428} = 91.82$. Check: 91.82 is about 92, and $(92)(92) = 8{,}464$ which is close to 8,428. So it looks all right.

ASSIGNMENT

My round number plant is real cute
It looks quite a bit like a jute
It grows very well
No one could tell
The poor thing has got a square root.

Find the following square roots.

1 $\sqrt{10.3}$
2 $\sqrt{27.4}$
3 $\sqrt{.00318}$
4 $\sqrt{.000562}$
5 $\sqrt{892{,}000.}$
6 $\sqrt{7{,}100.}$
7 $\sqrt{60}$
8 $\sqrt{.009}$
9 $\sqrt{51284.}$
10 $\sqrt{.00002419}$
11 $\sqrt{2{,}600}$
12 $\sqrt{260}$
13 $\sqrt{26}$
14 $\sqrt{2.6}$
15 $\sqrt{.26}$
16 $\sqrt{.026}$
17 $\sqrt{.0026}$
18 $\sqrt{.00026}$

TABLE B-3a A SHORT TABLE OF AREAS UNDER A NORMAL CURVE TO THE LEFT OF z

z Score	Proportion of Area to the Left of z
−4	.00003
−3	.001
−2.58	.005
−2.33	.01
−2	.023
−1.96	.025
−1.65	.05
−1	.16
0	.50
1	.84
1.65	.95
1.96	.975
2	.977
2.33	.99
2.58	.995
3	.999
4	.99997

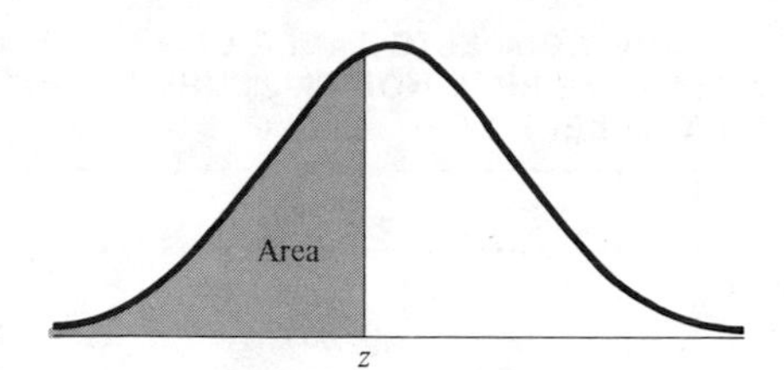

TABLE B-3b A MORE COMPLETE TABLE OF AREAS UNDER A NORMAL CURVE TO THE LEFT OF z

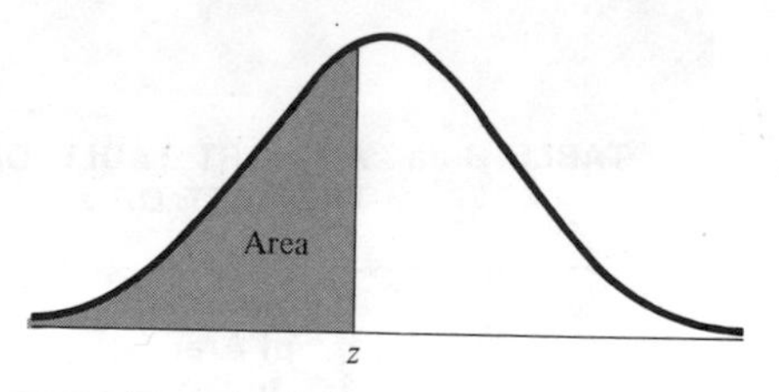

z	Area	z	Area	z	Area
−4	.00003	−2.74	.0031	−2.29	.0110
−3.9	.00005	−2.73	.0032	−2.28	.0113
−3.8	.0001	−2.72	.0033	−2.27	.0116
−3.7	.0001	−2.71	.0034	−2.26	.0119
−3.6	.0002	−2.70	.0035	−2.25	.0122
−3.5	.0002	−2.69	.0036	−2.24	.0125
−3.4	.0003	−2.68	.0037	−2.23	.0129
−3.3	.0005	−2.67	.0038	−2.22	.0132
−3.2	.0007	−2.66	.0039	−2.21	.0136
−3.1	.0010	−2.65	.0040	−2.20	.0139
−3.09	.0010	−2.64	.0041	−2.19	.0143
−3.08	.0010	−2.63	.0043	−2.18	.0146
−3.07	.0011	−2.62	.0044	−2.17	.0150
−3.06	.0011	−2.61	.0045	−2.16	.0154
−3.05	.0011	−2.60	.0047	−2.15	.0158
−3.04	.0012	−2.59	.0048	−2.14	.0162
−3.03	.0012	−2.58	.0049	−2.13	.0166
−3.02	.0013	−2.57	.0051	−2.12	.0170
−3.01	.0013	−2.56	.0052	−2.11	.0174
−3.00	.0013	−2.55	.0054	−2.10	.0179
−2.99	.0014	−2.54	.0055	−2.09	.0183
−2.98	.0014	−2.53	.0057	−2.08	.0188
−2.97	.0015	−2.52	.0059	−2.07	.0192
−2.96	.0015	−2.51	.0060	−2.06	.0197
−2.95	.0016	−2.50	.0062	−2.05	.0202
−2.94	.0016	−2.49	.0064	−2.04	.0207
−2.93	.0017	−2.48	.0066	−2.03	.0212
−2.92	.0017	−2.47	.0068	−2.02	.0217
−2.91	.0018	−2.46	.0069	−2.01	.0222
−2.90	.0019	−2.45	.0071	−2.00	.0228
−2.89	.0019	−2.44	.0073	−1.99	.0233
−2.88	.0020	−2.43	.0075	−1.98	.0239
−2.87	.0021	−2.42	.0078	−1.97	.0244
−2.86	.0021	−2.41	.0080	−1.96	.0250
−2.85	.0022	−2.40	.0082	−1.95	.0256
−2.84	.0023	−2.39	.0084	−1.94	.0262
−2.83	.0023	−2.38	.0087	−1.93	.0268
−2.82	.0024	−2.37	.0089	−1.92	.0274
−2.81	.0025	−2.36	.0091	−1.91	.0281
−2.80	.0026	−2.35	.0094	−1.90	.0287
−2.79	.0026	−2.34	.0096	−1.89	.0294
−2.78	.0027	−2.33	.0099	−1.88	.0301
−2.77	.0028	−2.32	.0102	−1.87	.0307
−2.76	.0029	−2.31	.0104	−1.86	.0314
−2.75	.0030	−2.30	.0107	−1.85	.0322

TABLE B-3b (*Continued*)

z	Area	z	Area	z	Area
−1.84	.0329	−1.39	.0823	− .94	.1736
−1.83	.0336	−1.38	.0838	− .93	.1762
−1.82	.0344	−1.37	.0853	− .92	.1788
−1.81	.0352	−1.36	.0869	− .91	.1814
−1.80	.0359	−1.35	.0885	− .90	.1841
−1.79	.0367	−1.34	.0901	− .89	.1867
−1.78	.0375	−1.33	.0918	− .88	.1894
−1.77	.0384	−1.32	.0934	− .87	.1922
−1.76	.0392	−1.31	.0951	− .86	.1949
−1.75	.0401	−1.30	.0968	− .85	.1977
−1.74	.0409	−1.29	.0985	− .84	.2005
−1.73	.0418	−1.28	.1003	− .83	.2033
−1.72	.0427	−1.27	.1020	− .82	.2061
−1.71	.0436	−1.26	.1038	− .81	.2090
−1.70	.0446	−1.25	.1056	− .80	.2119
−1.69	.0455	−1.24	.1075	− .79	.2148
−1.68	.0465	−1.23	.1093	− .78	.2177
−1.67	.0475	−1.22	.1112	− .77	.2206
−1.66	.0485	−1.21	.1131	− .76	.2236
−1.65	.0495	−1.20	.1151	− .75	.2266
−1.64	.0505	−1.19	.1170	− .74	.2296
−1.63	.0516	−1.18	.1190	− .73	.2327
−1.62	.0526	−1.17	.1210	− .72	.2358
−1.61	.0537	−1.16	.1230	− .71	.2389
−1.60	.0548	−1.15	.1251	− .70	.2420
−1.59	.0559	1.14	.1271	− .69	.2451
−1.58	.0571	−1.13	.1292	− .68	.2483
−1.57	.0582	−1.12	.1314	− .67	.2514
−1.56	.0594	−1.11	.1335	− .66	.2546
−1.55	.0606	−1.10	.1357	− .65	.2578
−1.54	.0618	−1.09	.1379	− .64	.2611
−1.53	.0630	−1.08	.1401	− .63	.2643
−1.52	.0643	−1.07	.1423	− .62	.2676
−1.51	.0655	−1.06	.1446	− .61	.2709
−1.50	.0668	−1.05	.1469	− .60	.2743
−1.49	.0681	−1.04	.1492	− .59	.2776
−1.48	.0694	−1.03	.1515	− .58	.2810
−1.47	.0708	−1.02	.1539	− .57	.2843
−1.46	.0722	−1.01	.1562	− .56	.2877
−1.45	.0735	−1.00	.1587	− .55	.2912
−1.44	.0749	− .99	.1611	− .54	.2946
−1.43	.0764	− .98	.1635	− .53	.2981
−1.42	.0778	− .97	.1660	− .52	.3015
−1.41	.0793	− .96	.1685	− .51	.3050
−1.40	.0808	− .95	.1711	− .50	.3085

TABLE B-3b (*Continued*)

z	Area	z	Area	z	Area
− .49	.3121	− .04	.4840	.41	.6591
− .48	.3156	− .03	.4880	.42	.6628
− .47	.3192	− .02	.4920	.43	.6664
− .46	.3228	− .01	.4960	.44	.6700
− .45	.3264	.00	.5000	.45	.6736
− .44	.3300	.01	.5040	.46	.6772
− .43	.3336	.02	.5080	.47	.6808
− .42	.3372	.03	.5120	.48	.6844
− .41	.3409	.04	.5160	.49	.6849
− .40	.3446	.05	.5199	.50	.6915
− .39	.3483	.06	.5239	.51	.6950
− .38	.3520	.07	.5279	.52	.6985
− .37	.3557	.08	.5319	.53	.7019
− .36	.3594	.09	.5359	.54	.7054
− .35	.3632	.10	.5398	.55	.7088
− .34	.3669	.11	.5438	.56	.7123
− .33	.3707	.12	.5478	.57	.7157
− .32	.3745	.13	.5517	.58	.7190
− .31	.3783	.14	.5557	.59	.7224
− .30	.3821	.15	.5596	.60	.7257
− .29	.3859	.16	.5636	.61	.7291
− .28	.3897	.17	.5675	.62	.7324
− .27	.3936	.18	.5714	.63	.7357
− .26	.3974	.19	.5753	.64	.7389
− .25	.4013	.20	.5793	.65	.7422
− .24	.4052	.21	.5832	.66	.7454
− .23	.4090	.22	.5871	.67	.7486
− .22	.4129	.23	.5910	.68	.7517
− .21	.4168	.24	.5948	.69	.7549
− .20	.4207	.25	.5987	.70	.7580
− .19	.4247	.26	.6026	.71	.7611
− .18	.4286	.27	.6064	.72	.7642
− .17	.4325	.28	.6103	.73	.7673
− .16	.4364	.29	.6141	.74	.7704
− .15	.4404	.30	.6179	.75	.7734
− .14	.4443	.31	.6217	.76	.7764
− .13	.4483	.32	.6255	.77	.7794
− .12	.4522	.33	.6293	.78	.7823
− .11	.4562	.34	.6331	.79	.7852
− .10	.4602	.35	.6368	.80	.7881
− .09	.4641	.36	.6406	.81	.7910
− .08	.4681	.37	.6443	.82	.7939
− .07	.4721	.38	.6480	.83	.7967
− .06	.4761	.39	.6517	.84	.7995
− .05	.4801	.40	.6554	.85	.8023

TABLE B-3b (*Continued*)

z	Area	z	Area	z	Area
.86	.8051	1.31	.9049	1.76	.9608
.87	.8078	1.32	.9066	1.77	.9616
.88	.8106	1.33	.9082	1.78	.9625
.89	.8133	1.34	.9099	1.79	.9633
.90	.8159	1.35	.9115	1.80	.9641
.91	.8186	1.36	.9131	1.81	.9649
.92	.8212	1.37	.9147	1.82	.9656
.93	.8238	1.38	.9162	1.83	.9664
.94	.8264	1.39	.9177	1.84	.9671
.95	.8289	1.40	.9192	1.85	.9678
.96	.8315	1.41	.9207	1.86	.9686
.97	.8340	1.42	.9222	1.87	.9693
.98	.8365	1.43	.9236	1.88	.9699
.99	.8389	1.44	.9251	1.89	.9706
1.00	.8413	1.45	.9265	1.90	.9713
1.01	.8438	1.46	.9278	1.91	.9719
1.02	.8461	1.47	.9292	1.92	.9726
1.03	.8485	1.48	.9306	1.93	.9732
1.04	.8508	1.49	.9319	1.94	.9738
1.05	.8531	1.50	.9332	1.95	.9744
1.06	.8554	1.51	.9345	1.96	.9750
1.07	.8577	1.52	.9357	1.97	.9756
1.08	.8599	1.53	.9370	1.98	.9761
1.09	.8621	1.54	.9382	1.99	.9767
1.10	.8643	1.55	.9394	2.00	.9772
1.11	.8665	1.56	.9406	2.01	.9778
1.12	.8686	1.57	.9418	2.02	.9783
1.13	.8708	1.58	.9429	2.03	.9788
1.14	.8729	1.59	.9441	2.04	.9793
1.15	.8749	1.60	.9452	2.05	.9798
1.16	.8770	1.61	.9463	2.06	.9803
1.17	.8790	1.62	.9474	2.07	.9808
1.18	.8810	1.63	.9484	2.08	.9812
1.19	.8830	1.64	.9495	2.09	.9817
1.20	.8849	1.65	.9505	2.10	.9821
1.21	.8869	1.66	.9515	2.11	.9826
1.22	.8888	1.67	.9525	2.12	.9830
1.23	.8907	1.68	.9535	2.13	.9834
1.24	.8925	1.69	.9545	2.14	.9838
1.25	.8944	1.70	.9554	2.15	.9842
1.26	.8962	1.71	.9564	2.16	.9846
1.27	.8980	1.72	.9573	2.17	.9850
1.28	.8997	1.73	.9582	2.18	.9854
1.29	.9015	1.74	.9591	2.19	.9857
1.30	.9032	1.75	.9599	2.20	.9861

TABLE B-3b (*Continued*)

z	Area	z	Area	z	Area
2.21	.9864	2.66	.9961	3.2	.9993
2.22	.9868	2.67	.9962	3.3	.9995
2.23	.9871	2.68	.9963	3.4	.9997
2.24	.9875	2.69	.9964	3.5	.9998
2.25	.9878	2.70	.9965	3.6	.9998
2.26	.9881	2.71	.9966	3.7	.9999
2.27	.9884	2.72	.9967	3.8	.9999
2.28	.9887	2.73	.9968	3.9	.99995
2.29	.9890	2.74	.9969	4.0	.99997
2.30	.9893	2.75	.9970		
2.31	.9896	2.76	.9971		
2.32	.9898	2.77	.9972		
2.33	.9901	2.78	.9973		
2.34	.9904	2.79	.9974		
2.35	.9906	2.80	.9974		
2.36	.9909	2.81	.9975		
2.37	.9911	2.82	.9976		
2.38	.9913	2.83	.9977		
2.39	.9916	2.84	.9977		
2.40	.9918	2.85	.9978		
2.41	.9920	2.86	.9979		
2.42	.9922	2.87	.9979		
2.43	.9925	2.88	.9980		
2.44	.9927	2.89	.9981		
2.45	.9929	2.90	.9981		
2.46	.9931	2.91	.9982		
2.47	.9932	2.92	.9982		
2.48	.9934	2.93	.9983		
2.49	.9936	2.94	.9984		
2.50	.9938	2.95	.9984		
2.51	.9940	2.96	.9985		
2.52	.9941	2.97	.9985		
2.53	.9943	2.98	.9986		
2.54	.9945	2.99	.9986		
2.55	.9946	3.00	.9987		
2.56	.9948	3.01	.9987		
2.57	.9949	3.02	.9987		
2.58	.9951	3.03	.9988		
2.59	.9952	3.04	.9988		
2.60	.9953	3.05	.9989		
2.61	.9955	3.06	.9989		
2.62	.9956	3.07	.9989		
2.63	.9957	3.08	.9990		
2.64	.9959	3.09	.9990		
2.65	.9960	3.10	.9990		

TABLE B-4 CRITICAL VALUES OF t FOR A TWO-TAILED TEST

(Values of t_c in this table are given without signs. All values are used as both positive and negative, e.g., $t_c = \pm 12.71$.)

Degrees of Freedom	t_c for $\alpha = .05$	t_c for $\alpha = .01$
1	12.71	63.66
2	4.30	9.92
3	3.18	5.84
4	2.78	4.60
5	2.57	4.03
6	2.45	3.71
7	2.36	3.50
8	2.31	3.36
9	2.26	3.25
10	2.23	3.17
11	2.20	3.11
12	2.18	3.06
13	2.16	3.01
14	2.14	2.98
15	2.13	2.95
16	2.12	2.92
17	2.11	2.90
18	2.10	2.88
19	2.09	2.86
20	2.09	2.84
21	2.08	2.83
22	2.07	2.82
23	2.07	2.81
24	2.06	2.80
25	2.06	2.79
26	2.06	2.78
27	2.05	2.77
28	2.05	2.76
29	2.04	2.76
30	2.04	2.75
40	2.02	2.70
50	2.01	2.68
60	2.00	2.66
80	1.99	2.64
100	1.98	2.63
120	1.98	2.62
200	1.97	2.60
500	1.96	2.59
Infinity	1.96	2.58

TABLE B-5 CRITICAL VALUES OF t FOR A ONE-TAILED TEST

(Values of t_c are given without signs. You must determine whether t_c is positive or negative from the alternative hypothesis.)

Degrees of Freedom	t_c for $\alpha = .05$	t_c for $\alpha = .01$
1	6.31	31.82
2	2.92	6.96
3	2.35	4.54
4	2.13	3.75
5	2.02	3.36
6	1.94	3.14
7	1.90	3.00
8	1.86	2.90
9	1.83	2.82
10	1.81	2.76
11	1.80	2.72
12	1.78	2.68
13	1.77	2.65
14	1.76	2.62
15	1.75	2.60
16	1.75	2.58
17	1.74	2.57
18	1.73	2.55
19	1.73	2.54
20	1.72	2.53
21	1.72	2.52
22	1.72	2.51
23	1.71	2.50
24	1.71	2.49
25	1.71	2.48
26	1.71	2.48
27	1.70	2.47
28	1.70	2.47
29	1.70	2.46
30	1.70	2.46
40	1.68	2.42
50	1.68	2.40
60	1.67	2.40
80	1.66	2.37
100	1.66	2.36
120	1.66	2.36
200	1.65	2.34
500	1.65	2.33
Infinity	1.65	2.33

TABLE B-6 CRITICAL VALUES OF THE COEFFICIENT OF CORRELATION r_c FOR A TWO-TAILED TEST

(Values of r are given in this table without signs. All values are both positive and negative, e.g., $r_c = \pm 1.00$.)

n	r_c for $\alpha = .05$	r_c for $\alpha = .01$
3	1.00	1.00
4	.95	.99
5	.88	.96
6	.81	.92
7	.75	.87
8	.71	.83
9	.67	.80
10	.63	.76
11	.60	.73
12	.58	.71
13	.53	.68
14	.53	.66
15	.51	.64
16	.50	.61
17	.48	.61
18	.47	.59
19	.46	.58
20	.44	.56
21	.43	.55
22	.42	.54
23	.41	.53
24	.40	.52
25	.40	.51
26	.39	.50
27	.38	.49
28	.37	.48
29	.37	.47
30	.36	.46

For values of r_c, when n is greater than 30 use

$$r_c = \frac{t_c}{\sqrt{t_c^2 + (n-2)}}$$

where t_c is the corresponding critical value of t for $(n-2)$ degrees of freedom in Table B-4.

TABLE B-7 CRITICAL VALUES OF THE COEFFICIENT OF CORRELATION r_c FOR A ONE-TAILED TEST

(Values of r are given in this table without signs. You must determine whether the critical values of r are positive or negative from the alternative hypothesis.)

n	r_c for $\alpha = .05$	r_c for $\alpha = .01$
3	.99	1.00
4	.90	.98
5	.81	.93
6	.73	.88
7	.67	.83
8	.62	.79
9	.58	.75
10	.54	.72
11	.52	.69
12	.50	.66
13	.48	.63
14	.46	.61
15	.44	.59
16	.42	.57
17	.41	.56
18	.40	.54
19	.39	.53
20	.38	.52
21	.37	.50
22	.36	.49
23	.35	.48
24	.34	.47
25	.34	.46
26	.33	.45
27	.32	.45
28	.32	.44
29	.31	.43
30	.31	.42

For values of r_c, when n is greater than 30 use

$$r_c = \frac{t_c}{\sqrt{t_c^2 + (n-2)}}$$

where t_c is the corresponding critical value of t for $(n-2)$ degrees of freedom in Table B-5.

TABLE B-8 CRITICAL VALUES FOR RANK SUM TEST

(The values in the table are the critical values for W, the rank sum of the smaller size sample.)

Sample Sizes (Smaller, Larger)	W_c for $\alpha = .01$	W_c for $\alpha = .05$
5, 5	15, 40	17, 38
5, 6	16, 44	18, 42
5, 7	17, 48	20, 45
5, 8	18, 52	21, 49
5, 9	18, 57	22, 53
5, 10	19, 61	23, 57
6, 6	23, 55	26, 52
6, 7	24, 60	27, 57
6, 8	25, 65	29, 61
6, 9	26, 70	31, 65
6, 10	28, 74	32, 70
7, 7	32, 73	36, 69
7, 8	34, 78	38, 74
7, 9	35, 84	40, 79
7, 10	37, 89	42, 84
8, 8	44, 92	49, 87
8, 9	45, 99	51, 93
8, 10	47, 105	53, 99
9, 9	57, 114	63, 108
9, 10	59, 121	65, 115
10, 10	71, 139	78, 132

For large n_1, n_2 ($n_2 \geq n_1 > 10$)

$$W_c = \mu_W \pm z_c \sigma_n$$

where $$\mu_W = \frac{n_1(n_1 + n_2 + 1)}{2}$$

$$\sigma_W = \sqrt{\frac{n_1 n_2 (n_1 + n_2 + 1)}{12}}$$

TABLE B-9 CRITICAL VALUES FOR RANK CORRELATION COEFFICIENT r_s

Number of Pairs n	r_S for $\alpha = .05$
6	−.77, .83
7	−.71, .75
8	−.69, .71
9	−.67, .68
10	−.62, .64

If $n > 10$, we can find the critical values from

$$r_{s_c} = \frac{\pm Z_c}{\sqrt{n-1}}$$

Example: $n = 11$, $\alpha = .05$
Then

$$r_c = \frac{+1.96}{\sqrt{11-1}} = \frac{\pm 1.96}{\sqrt{10}} = \pm\frac{1.96}{3.16}$$

$$= \pm .62$$

APPENDIX C:

Formulas

FORMULAS:

Chapter 3

1 $P(\text{event}) = \dfrac{\text{number of favorable outcomes}}{\text{total number of equally likely outcomes}} = \dfrac{f}{n}$

2 $q = 1 - p$

Chapter 4

1 $E = \dfrac{(\text{row total})(\text{column total})}{n} = \dfrac{(RT)(CT)}{n}$

2 a $\text{df} = C - 1$, if only one row
  b $\text{df} = (R-1)(C-1)$, if more than one row

3 Outcome: $X^2 = \Sigma \left(\dfrac{O^2}{E}\right) - n$

Chapter 5

1 $\mu_p = \pi$

2 $\sigma_p = \sqrt{\dfrac{\pi(1-\pi)}{n}}$

3 $z_X = \dfrac{X - \mu}{\sigma}$

4 $X = \mu + z\sigma$

5 $p_c = \mu_p + z_c\sigma_p$

6 Outcome $= p$

Chapter 6

1 $p_1 = \dfrac{f_1}{n_1}$

2 $p_2 = \dfrac{f_2}{n_2}$

3 $p = \dfrac{f_1 + f_2}{n_1 + n_2}$

4 $dp = p_1 - p_2$

5 $\mu_{dp} = \pi_1 - \pi_2$

6 $s_{dp} = \sqrt{\dfrac{pq}{n_1} + \dfrac{pq}{n_2}}$

7 $dp_c = \mu_{dp} + z_c s_{dp}$

8 Outcome $= p_1 - p_2$

Chapter 7

1 $\mu = \dfrac{\Sigma X}{n}$

3 $s_{\text{pop}} = \sqrt{\dfrac{\Sigma X^2 - \dfrac{(\Sigma X)^2}{n}}{n-1}}$

4 $\sigma_m = \dfrac{\sigma_{\text{pop}}}{\sqrt{n}}$ $\quad s_m = \dfrac{s_{\text{pop}}}{\sqrt{n}}$

5 a $m_c = \mu_m + z_c\, \sigma_m,\ n > 30$
  b $m_c = \mu_m + t_c\, \sigma_m,\ n \ngtr 30$

6 $\text{df} = n - 1$

7 Outcome $= m$

Chapter 8

1 $dm = m_1 - m_2$

2 $\mu_{dm} = \mu_1 - \mu_2$

3 $\sigma_{dm} = \sqrt{\frac{\sigma_1^2}{n_1} + \frac{\sigma_2^2}{n_2}} \qquad s_{dm} = \sqrt{\frac{s_1^2}{n_1} + \frac{s_2^2}{n_2}}$

4 a $dm_c = \mu_{dm} + z_c s_{dm}$, $\quad (n_1 \text{ and } n_2 > 30)$

b $dm_c = \mu_{dm} + t_c s_{dm}$, $\quad (n_1 \text{ or } n_2 \not> 30)$

5 $\text{df} = n_1 + n_2 - 2$

6 $\text{Outcome} = m_1 - m_2$

Chapter 9

1 $s_p = \sqrt{\frac{pq}{n}}$

2 $p - z_c s < \pi < p + z_c s$

3 a $s_{dp} = \sqrt{\frac{p_1 q_1}{n_1} + \frac{p_2 q_2}{n_2}}$

b $(p_1 - p_2) - z_c s_{dp} < \pi_1 - \pi_2 < (p_1 - p_2) + z_c s_{dp}$

4 a $m - z_c s_m < \mu_{\text{pop}} < m + z_c s_m$, $\quad (n > 30)$

b $m - t_c s_m < \mu_{\text{pop}} < m + t_c s_m$, $\quad (n \not> 30)$

5 a $(m_1 - m_2) - z_c s_{dm} < \mu_1 - \mu_2 < (m_1 - m_2) + z_c s_{dm} \quad (n_1 \text{ and } n_2 > 30)$

b $(m_1 - m_2) - t_c s_{dm} < \mu_1 - \mu_2 < (m_1 - m_2) + t_c s_{dm} \quad (n_1 \text{ or } n_2 \not> 30)$

Chapter 10

1 $r = \frac{n\Sigma XY - (\Sigma X)(\Sigma Y)}{\sqrt{n\Sigma X^2 - (\Sigma X)^2}\ \sqrt{n\Sigma Y^2 - (\Sigma Y)^2}}$

2 $b = \frac{n\Sigma XY - (\Sigma X)(\Sigma Y)}{n\Sigma X^2 - (\Sigma X)^2}$

3 $\tilde{Y} = b(X - m_X) + m_Y$

Chapter 11

1 $\mu_W = \frac{n_1(n_1 + n_2 + 1)}{2} \quad (n_1 \leq n_2)$

2 $\sigma_W = \sqrt{\frac{n_1 n_2 (n_1 + n_2 + 1)}{12}}$

3 $W_c = \mu_W + z_c \sigma_W$

4 $S = \Sigma(\text{differences})^2$

5 $r_s = 1 - \frac{6S}{n(n^2 - 1)} \quad (n \leq 10)$

6 $r_s = \frac{\pm z_c}{\sqrt{n - 1}} \quad (n > 10)$

A SUMMARY OF HYPOTHESIS TESTS

Name of Test	When Used	Conditions	Formulas
I Chi Square	Are two variables related?	Each $E > 5$. Not 2×2. Use chi-square distribution	1 $E = \frac{(CT)(RT)}{n}$ 2 $X^2 = \Sigma\left(\frac{O^2}{E}\right) - n$ 3 $\text{df} = C - 1$ if $R = 1$ $\text{df} = (R-1)(C-1)$ if $R > 1$
II Binomial One-Sample	Is a claim about a proportion correct?	$n\pi$ and $n(1-\pi) > 5$. Use normal distribution	1 $\mu_p = \pi$ 2 $s_p = \sqrt{\frac{\pi(1-\pi)}{n}}$
III Binomial Two-Sample	Are two proportions equal?	Four outcomes all greater than 5. Use normal distribution	1 $\mu_{dp} = \pi_1 - \pi_2$ 2 $s_{dp} = \sqrt{\frac{pq}{n_1} + \frac{pq}{n_2}}$ 3 $dp = p_1 - p_2$
IV A Sample-Mean, One Large Sample	Is a claim about a mean correct?	$n > 30$. Use normal distribution	1 $\mu_m = \mu_{\text{pop}}$ 2 $s_{\text{pop}} = \sqrt{\frac{\Sigma X^2 - \frac{(\Sigma X)^2}{n}}{n-1}}$ 3 $s_m = \frac{s_{\text{pop}}}{\sqrt{n}}$
IV B Sample-Mean, One Small Sample	Is a claim about a mean correct?	$n \leq 30$ and population is normal. Use t distribution	1 $\mu_m = \mu_{\text{pop}}$ 2 $s_{\text{pop}} = \sqrt{\frac{\Sigma X^2 - \frac{(\Sigma X)^2}{n}}{n-1}}$ 3 $s_m = \frac{s_{\text{pop}}}{\sqrt{n}}$ 4 $\text{df} = n - 1$

A SUMMARY OF HYPOTHESIS TESTS *(Continued)*

Name of Test	When Used	Conditions	Formulas
V A Sample-Mean, Two Samples Large	Are two means equal?	n_1 and n_2 both greater than 30. Use normal distribution	**1** $\mu_{dm} = \mu_1 - \mu_2$ **2** $s_{dm} = \sqrt{\frac{s_1^2}{n_1} + \frac{s_2^2}{n_2}}$
V B Sample-Mean, Two Samples Small	Are two means equal?	Populations normal and n_1 or $n_2 \ngtr 30$. Use t distribution	**1** $\mu_{dm} = \mu_1 - \mu_2$ **2** $s_{dm} = \sqrt{\frac{s_1^2}{n_1} + \frac{s_2^2}{n_2}}$ **3** $\text{df} = n_1 + n_2 - 2$
VI Linear correlation	Are two variables related linearly?	Populations normal $n > 2$	**1** $r = \frac{n\Sigma XY - (\Sigma X)(\Sigma Y)}{\sqrt{n\Sigma X^2 - (\Sigma X)^2}\,\sqrt{n\Sigma Y^2 - (\Sigma Y)^2}}$
VII Rank Sum	Are two populations equally ranked?	n_1 and $n_2 \geq 5$ $[n_1 \leq n_2]$	**1** $\mu_W = \frac{n_1(n_1 + n_2 + 1)}{2}$ **2** $\sigma_W = \sqrt{\frac{n_1 n_2(n_1 + n_2 + 1)}{12}}$
VIII Rank Correlation	Are two sets of rankings related linearly?	$n \geq 6$	**1** $r_s = 1 - \frac{6S}{n(n^2 - 1)}$ **2** $r_c = \pm \frac{z_c}{\sqrt{n - 1}}$

ESTIMATION SUMMARY

	Parameter	Point Estimate	Conditions	Confidence Interval	Formulas
I Binomial One-Sample	π	p	Two outcomes greater than 5	$p - zs_p < \pi < p + zs_p$	$s_p = \sqrt{\frac{pq}{n}}$ [°]
II Binomial Two-Sample	$\pi_1 - \pi_2$	$p_1 - p_2$	Four outcomes greater than 5	$(p_1 - p_2) - zs_{dp} < \pi_1 - \pi_2 < (p_1 - p_2) + zs_{dp}$	$s_{dp} = \sqrt{\frac{p_1q_1}{n_1} + \frac{p_2q_2}{n_2}}$ [°]
III A Sample-Mean One-Sample, Large Sample	μ	m	$n > 30$	$m - zs_m < \mu_{pop} < m + zs_m$	$s_m = \frac{s_{pop}}{\sqrt{n}}$
III B Sample-Mean One-Sample, Small Sample	μ	m	$n \not> 30$ population is normal	$m - ts_m < \mu_{pop} < m + ts_m$	$df = n - 1$
IV A Sample-Mean Two-Sample, Large Sample	$\mu_1 - \mu_2$	$m_1 - m_2$	$n_1 > 30$ and $n_2 > 30$	$(m_1 - m_2) - zs_{dm} < \mu_1 - \mu_2 < (m_1 - m_2) + zs_{dm}$	$s_{sm} = \sqrt{\frac{s_1^2}{n_1} + \frac{s_2^2}{n_2}}$
IV B Sample-Mean Two-Sample, Small Sample	$\mu_1 - \mu_2$	$m_1 - m_2$	$n_1 \not> 30$ or $n_2 > 30$ both populations are normal	$(m_1 - m_2) - ts_{dm} < \mu_1 - \mu_2 < (m_1 - m_2) + ts_{dm}$	$df = n_1 + n_2 - 2$

[°] These two formulas for binomial confidence intervals are different from the ones used for hypothesis testing.

APPENDIX D:

One-Tail Hypothesis Tests

In the regular chapters of this text we discuss hypothesis tests by stating a null hypothesis and an alternative hypothesis. In all the examples the null hypothesis has the form H_0: Some population parameter *is equal to* some value.

For example, $H_0: \mu = 100$

The corresponding alternative hypothesis has the form H_a: Some population parameter *is not equal to* some value.

For example, $H_a: \mu \neq 100$

Then we observe the data and use some statistic based on this sample data to decide if H_0 is likely to be false.

A null hypothesis and its corresponding alternative are phrased so that if one is true then the other is false. There are other possible types of hypotheses besides the two we have shown so far.

Here is an illustration of how more than these two hypotheses might happen. Suppose an engineer has invented a new design for an artificial heart. Suppose she wants to demonstrate that it is *superior* to the old type. In order to do this she will have to produce data which show, for instance, that her design lasts *longer* than the old kind. No one will be interested if it lasts for a shorter time than the old kind. So, from the statistical point of view, she wants to show not just that her design is *different* from the old one, but that it is different *in a particular direction,* namely the direction of lasting *longer*.

She would establish the following pair of statistical hypotheses:

H_0: The new type of heart *is no better than* the old one.
H_a: The new type of heart *is better than* the old one.

To make the problem more specific suppose that the old kind of artificial heart lasted an average of 600 days. So she wants to demonstrate that her invention lasts more than 600 days. This means that if she tests a *sample* of new hearts, they will have to last significantly longer than 600 days; enough longer so that we will be pretty sure we did not get a few favorable results just by luck.

Her hypotheses would be these:

H_0: The average life of the new heart is not longer than 600 days.
H_a: The average life of the new heart *is longer than* 600 days.

In symbols, these two hypotheses are written like this:

$H_0: \mu \leq 600$ ($\leq$ means "is less than or equal to" which is the same as "not more than")
$H_a: \mu > 600$ ($>$ means "is more than")

Let us see what effect this has on the rest of the testing procedure.

Because the engineer hopes to prove that her product lasts *longer* (and not shorter) than the old product, she will only be interested in showing that her samples have a mean life which is *more than* some critical value. In short, an alternative hypothesis symbolized by a "greater than" sign

(>) leads us to look only at *one* critical value instead of the usual two. A hypothesis test which leads to *one* critical value is called a **one-tail test.** A hypothesis test which leads to *two* critical values is called a **two-tail test.**

We will go through a problem now with enough data to do all the necessary computations.

Illustration

An engineer has designed a new type of artificial heart. The old type of heart has a mean life of 600 days with a standard deviation of 90 days. She is going to test a random sample of 36 of the new type. How long would the average life of the hearts in this sample have to be in order to claim that they are *better* than the old ones? Allow $\alpha = .05$, and assume σ is still 90.

Solution

1 Population: New type heart.
2 H_0 : Average life of new heart not longer than 600 days. $\mu \leq 600$.
H_a : Average life of new heart is longer than 600 days. $\mu > 600$.
3 We base our decision rule on the assumption that $\mu = 600$. Since $\alpha = .05$, we are willing to set our decision rule so that there is a .05 probability of a Type I error. But since this is a one-tail test, we will only attempt to find one critical value. Therefore, our sketch will look like Figure D-1. We have marked the critical value in the *right* tail of the curve because that side of the graph corresponds to life spans *higher* than 600.

We know that for n large, the distribution of sample means is normal with

$$\mu_m = \mu_{\text{pop}} \quad \text{and} \quad \sigma_m = \frac{\sigma_{\text{pop}}}{\sqrt{n}}$$

So we get $\mu_m = 600$ days, and $\sigma_m = 90/\sqrt{36} = 15$ days.

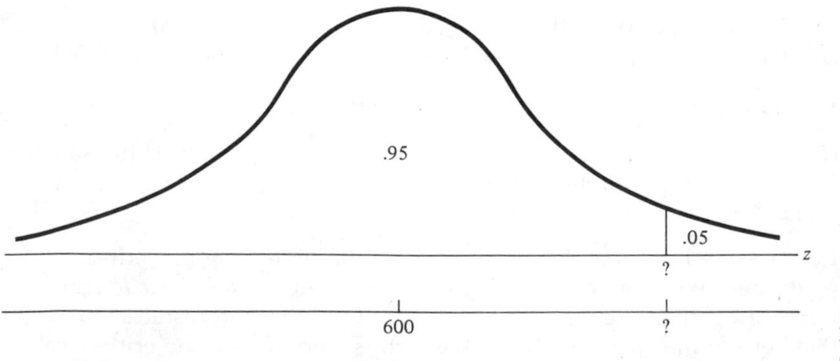

Figure D-1

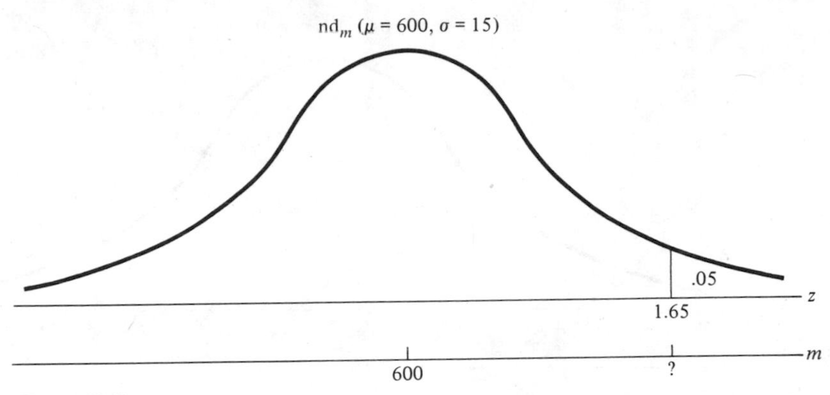

Figure D-2

Looking in the normal curve table, we find that the critical z score which cuts off the *right-hand 5 percent*, is $z_c = 1.65$. So our picture looks like Figure D-2.

We compute the critical value of m for the decision rule.

$$\begin{aligned} m_c &= \mu_m + z_c\sigma_m \\ &= 600 + (1.65)(15) \\ &= 600 + 24.75 \\ &= 625.75 \end{aligned}$$

Therefore, we arrive at the following decision rule: If the sample of new hearts has an average life more than 625.75 days, we will conclude that the new hearts last longer than the old hearts.

Second Illustration

A sales analyst develops a new marketing procedure which he feels will *reduce* the percentage of returned goods. The old procedure results in 15 percent of the goods being returned. If he is going to test his new procedure on a batch of 1,000 items, how small should the percentage of returned goods be in order to support the claim that the new procedure *reduces* returns? Assume he is willing to let $\alpha = .05$.

1 Population: Goods marketed under the new procedure.
2 Hypotheses: $H_0: \pi \geq .15$. The percentage of returns is not less than 15 percent.
  $H_a: \pi < .15$. The percentage of returns is less than 15 percent.
3 Decision Rule: Assume that $\pi = .15$. We have

$$n\pi = (1{,}000)(.15) = 150$$

and

$$n(1 - \pi) = (1{,}000)(.85) = 850$$

Both greater than 5.

Therefore, the distribution of sample values of p will be normal with

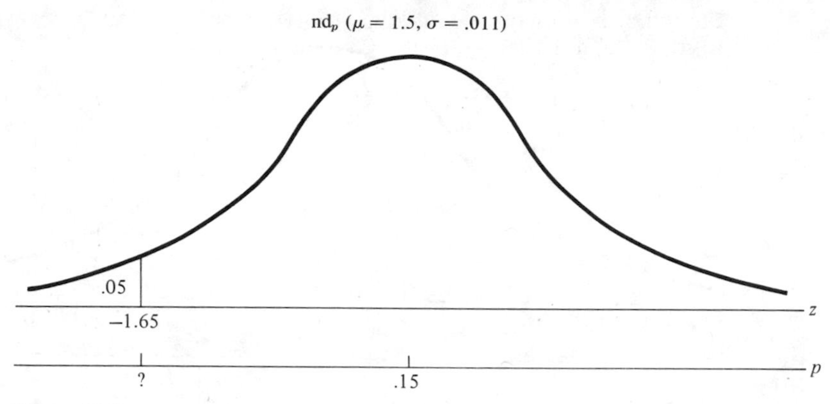

Figure D-3

$\mu_p = \pi$ and

$$\sigma_p = \sqrt{\frac{\pi(1-\pi)}{n}}$$

So $\mu_p = .15$ and $\sigma_p = \sqrt{\frac{(.15)(.85)}{1{,}000}} = \sqrt{.0001275} = .011$

We get the picture shown in Figure D-3.

Because we are interested only in *reducing* the percentage of returns, we have H_a symbolized using a "less than" sign (<). There, we have a one-tail test on the left.

Looking at the normal curve table, we see that the critical value of z is −1.65.

$$\begin{aligned} p_c &= \mu_p + \sigma_p z_c \\ &= .15 + (.011)(-1.65) \\ &= .15 - .018 \\ &= .132 \end{aligned}$$

Therefore, we make the following conclusion: If the marketing sample using the new procedure results in *less than* 13.2 percent returns, we will decide that the new method does reduce returns.

Third Illustration

(Two-sample comparison.) An old vitamin freak, Granny Ola (who is a statistician on the side), wants to show that taking vitamin G just before lunch is *superior* to taking vitamin G just after lunch. The test of effectiveness is the speed of eye blinking. She invites a random sample of 36 friends for a wheat germ sandwich and gives them each vitamin G before lunch. An hour later she measures how fast each person can blink. The next day she invites another random sample of 36 friends and gives them the vitamin after the sandwich. Again, an hour later she measures their blinking speed. Here are the tabulated results:

	Average Blinking Speed m	Standard Deviation s
Before lunch group	470 blinks per minute	120 blinks per minute
After lunch group	380 blinks per minute	100 blinks per minute

She wants to prove that the before lunch results are significantly different from the after lunch results and, in particular, that the before lunch results are significantly *higher* than the after lunch results. She is using $\alpha = .01$.

1 Populations
pop$_1$: People who take vitamin G before lunch.
pop$_2$: People who take vitamin G after lunch.
2 Hypotheses
H_0: The pop$_1$ blinking speed is no higher than the pop$_2$ blinking speed.
$\mu_1 - \mu_2 \leq 0$. $dm \leq 0$.
H_a: The pop$_1$ blinking speed *is higher* than the pop$_2$ blinking speed.
$\mu_1 - \mu_2 > 0$. $dm > 0$.
3 Decision Rule
Since n_1 and n_2 are both bigger than 30, we can assume that the dm's are distributed normally with mean 0 and

$$\text{standard deviation} = \sqrt{\frac{s_1^2}{n_1} + \frac{s_2^2}{n_2}}$$

$$s_{dm} = \sqrt{\frac{s_1^2}{n_1} + \frac{s_2^2}{n_2}} = \sqrt{\frac{120^2}{36} + \frac{100^2}{36}}$$

$$= \sqrt{\frac{14{,}400}{36} + \frac{10{,}000}{36}} = \sqrt{\frac{24{,}400}{36}}$$

See Figure D-4. Looking at the normal curve table for z_c with 99 percent to the left, we get $z_c = 2.33$.

$$\begin{aligned} dm_c &= \mu_{dm} + z_c s_{dm} \\ &= 0 + (2.33)(26.03) \\ &= 60.65 \end{aligned}$$

If the difference between m_1 and m_2 shows m_1 *higher* by more than 60.65 blinks per minute, we will say that the evidence indicates we should reject H_0. We will want to claim that Granny seems to be correct.
4 Examine the sample results. We have $m_1 - m_2 = 470 - 380 = 90$ blinks per minute. We conclude that Granny was correct. It seems that taking the vitamins before lunch is more effective than taking them after lunch.

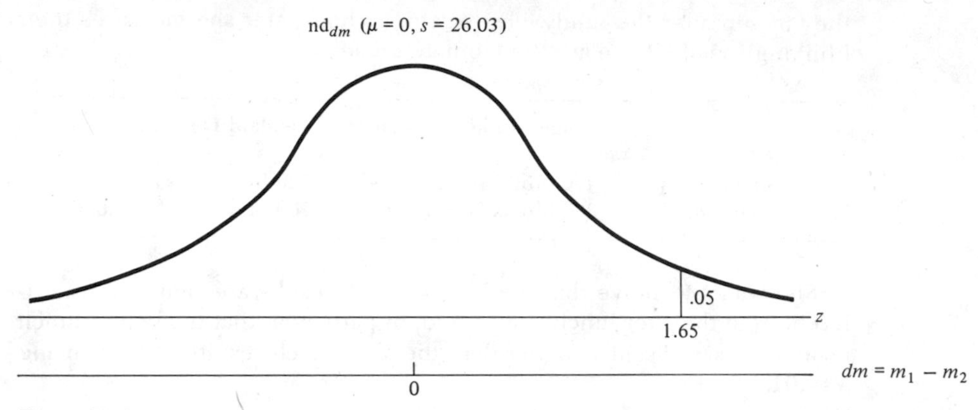

Figure D-4

Summary

1 If an alternative hypothesis is symbolized using a "greater than" (>) sign, we have a one-tail test on the right.
2 If it is symbolized with a "less than" (<) sign, we have a one-tail test on the left.
3 If it is symbolized with a "not equal to" (≠) sign, we have a two-tail test.

ASSIGNMENT

A test with only one tail
Was performed in the Dade County jail
Do more of the men
Wind up in the pen?
Why yes, said the girl out on bail.

Note to teacher: Go back to the assignments in previous chapters. Many of the hypothesis tests can be changed to one-tail tests.

APPENDIX E:

Type II Errors

TYPE II ERRORS

A Type II error occurs when the sample data lead a statistician to accept a *false* null hypothesis. The probability of this happening is denoted by the Greek letter β (beta). In general beta is not known, but it can often be estimated.

For example if the null hypothesis, "a coin is fair" is false, then $\pi = p(\text{heads}) \neq .5$. If $\pi = .6$, beta will have one value, but if $\pi = .9$, beta will have a different value. For each different value of π, we get a different value of beta.

Suppose we were testing this null hypothesis with $\alpha = .05$ and $n = 60$.

$H_0 : \pi = .5$
$H_a : \pi \neq .5$
$n = 60$
$n\pi$ and $\pi(1 - \pi)$ are both greater than 5 so we can use the normal curve.

$$\sigma_p = \sqrt{\frac{(.5)(.5)}{60}} = .065$$

Figure E-1 shows this graphically.

$$p_c = \mu_p + z_c\sigma_p$$
$$= .5 + (\pm 1.96)(.065) = .500 \pm .127 = .373 \text{ and } .627$$

Our decision rule would be: Reject the null hypothesis if the outcome is less than .373 or greater than .627. *If the null hypothesis is false,* then we would make a Type II error whenever we obtained an outcome between .373 and .627. The probability of this happening, beta, depends on the true value of π.

Case I

Suppose $\pi = .6$. The mean is no longer .5. Rather $\mu_p = .6$ and

$$\sigma_p = \sqrt{\frac{(.6)(.4)}{60}} = \sqrt{.004} = .063$$

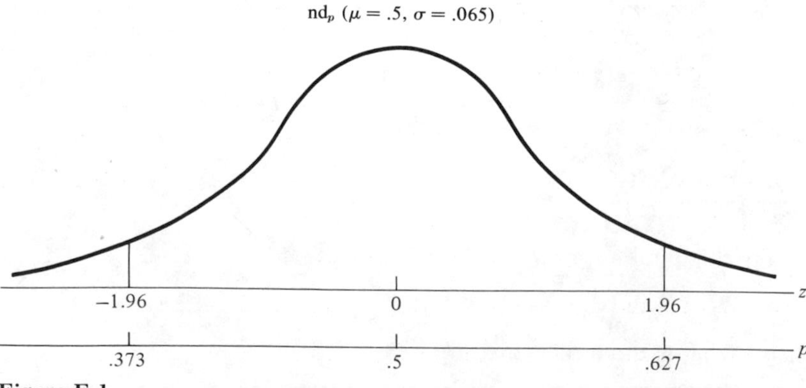

Figure E-1

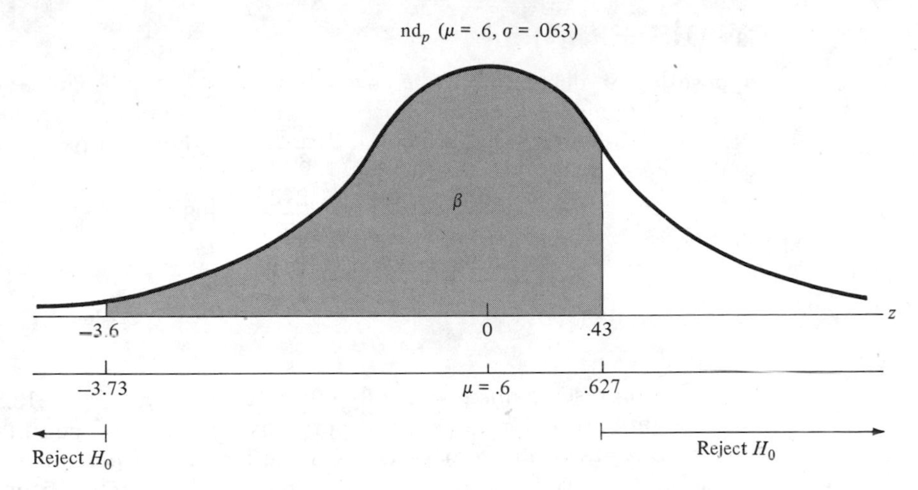

Figure E-2

$$z_{.627} = \frac{X - \mu}{\sigma} = \frac{.627 - .600}{.063} = \frac{.627 - .600}{.063} = .43$$

$$z_{.373} = \frac{.373 - .600}{.063} = \frac{-.227}{.063} = -3.6$$

See Figure E-2 for the graphic representation.

Now the probability of an outcome between $z = -3.6$ and $z = .43$ can be seen from Figure E-2 to be well over 50 percent. (Using Table B-36 it can be found, in fact, to be about 66 percent.) So we have that the probability of a Type II error is approximately 66 percent. What does this mean? It sounds bad to have the probability of any kind of error equal to 66 percent; and it is bad!

What does the 66 percent measure? It says that under certain conditions we will make the wrong decision 66 percent of the time, using our decision rule. What conditions are these? Recall that we are trying to decide which hypothesis ($\pi = .5$ or $\pi = .6$) is true on the basis of 60 observations. That β is so high reflects the fact that it is almost impossible to distinguish between these two hypotheses on the basis of so few observations. These two hypotheses are not very different from one another in the first place.

If you want to distinguish between two hypotheses which are close to each other, and if you want both α and β to be low, then you will have to have many more than 60 observations.

CLASS EXERCISE

E-1 Try this same experiment but take $n = 1{,}000$.

a What is the new decision rule? ______________________

b What is the new value of beta? ______________________

Case II

Suppose $\pi = .9$. This time we have

$$\mu_p = .9 \quad \text{and} \quad \sigma_p = \sqrt{\frac{(.9)(.1)}{60}} = \sqrt{.0015} = .04$$

$$z_{.627} = \frac{.627 - .900}{.04} = -\frac{.273}{.04} = -6.8$$

$$z_{.373} = \frac{.373 - .900}{.04} = -\frac{.527}{.04} = -13.2$$

Figure E-3 represents this graphically. Both $z = -6.8$ and $z = -13.2$ are so far to the left that for all practical purposes the area between them is zero. So β is practically zero, certainly much less than 1 percent.

What all this says is that if the coin is fair, we hav only a 5 percent chance of a Type I error occurring. If the coin is sligl ly biased with π about .6, we run a 66 percent chance of a Type II error. However, if the coin is very biased with π about .9, we run less than a 1 percent chance of a Type II error because the two hypotheses ($\pi = .5$ and $\pi = .9$) are quite different, and it should be easy to find out which one is true even with relatively few observations.

The value of beta depends on the true value of the parameter being investigated, the sample size, and the significance level. In general, increasing n will decrease β. Similarly, for any fixed value of n, increasing α will decrease β, and decreasing α will increase the value of β.

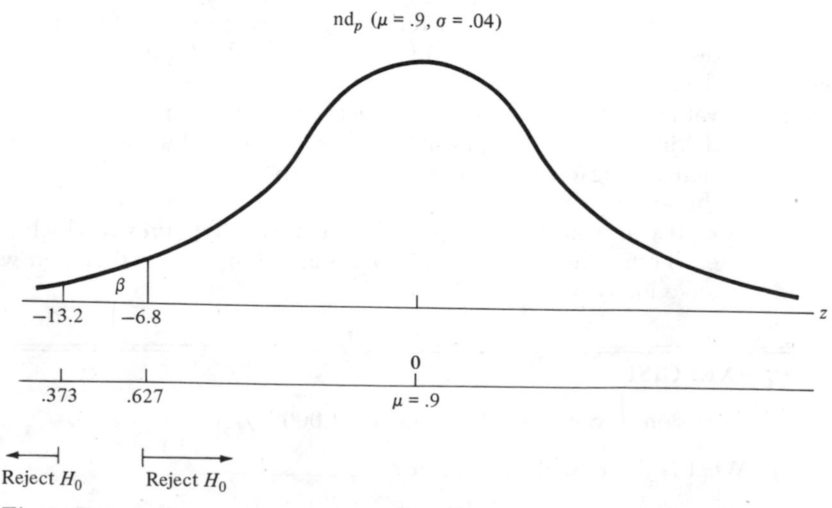

Figure E-3

ASSIGNMENT

Her candy was awful; I hate her
Said the young man from Decatur
I "proved" it's too sweet
By a trick very neat—
I admit I was fudging the β

1 A wheel is divided into quarters and painted red, yellow, blue, and green. Let $\pi = P(\text{red})$. If the wheel is fair, $\pi = .25$.
  a Find the critical values p_c if $\alpha = .05$ and $n = 200$.
  b Using the critical values found in part a, estimate the value of β if π is really .40.

2 Repeat assignment 1 with $\alpha = .01$.

3 A hypothesis test that the average weight of abominable snowmen is 6,500 pounds was performed. An average weight of 6,300 pounds and $s = 400$ pounds was found for 1,000 snowmen. Let $\alpha = .05$.
  a Find the critical values m_c.
  b Estimate β for the hypothesis test if μ is actually 6,000 pounds.

APPENDIX F:

Bibliography

Statistics Texts

Slightly more advanced:

1 Chapman, D., and R. Schaufele: "Elementary Probability Models and Statistical Inference," Ginn-Blaisdell, Lexington, Mass., 1970.
2 David, F. N.: *A First Course in Statistics,* Monograph 31 of Griffin's Statistical Monographs and Courses, Hafner, New York, 1971.
3 Dixon, W., and F. Massey: "Introduction to Statistical Analysis," McGraw-Hill, 1969.
4 Fisher, R. S.: "Statistical Method for Research Workers," 14th ed., Hafner, New York, 1973.
5 Freund, J. E.: "Modern Elementary Statistics," Prentice-Hall, Englewood Cliffs, N.J., 1967.
6 Mendenhall, W.: "Introduction to Probability and Statistics," Duxbury Press, Scituate, Mass., 1971.

You need to know some calculus for these:

7 Hoel, P., S. Port, and C. Stone: "Introduction to Probability Theory," Houghton Mifflin, Boston, Mass., 1971.
8 Hoel, P., S. Port, and C. Stone: "Introduction to Statistical Theory," Houghton Mifflin, Boston, Mass., 1971.

Related Readings

9 David, F. N.: "Games, Gods and Gambling," Charles Griffin & Company, London (also Hafner), 1962.
10 Fadiman, Clifton (ed.): "Fantasia Mathematica," Simon and Schuster, New York, 1958.
11 Huff, D.: "How To Lie With Statistics," W. W. Norton and Company, New York, 1954. "How To Take A Chance," W. W. Norton and Company, New York, 1959.
12 Slonim, M.: "Sampling in a Nutshell" (Also published under the title "Sampling, A Quick Reliable Guide to the Unknown"), Simon and Schuster, New York, 1966.
13 Tanur, J. (ed.): "Statistics: A Guide to the Unknown," Holden-Day, San Francisco Calif., 1972. Especially these readings which are of intrinsic interest but which do not fit in any particular chapter of this text.
Early Warning Signals for the Economy, Moore and Shiskin, p. 310. What are economic *indicators*?
How Crowded Will We Become?, Keyfitz, p. 297. How to estimate population size in the future.
Measuring the Effects of Social Innovations by Means of Time Series, Campbell, p. 120. What are *time-series*?

APPENDIX G:

Selected Readings

These three readings are illustrations of some of the ways that statistics are used to describe human behavior. None of the articles is very technical, but they have to be read carefully. After each article we have asked one or two questions to suggest what types of questions can be asked and answered using the ideas in this book. We hope that each article can serve as a focus for classroom discussion, and we recommend that they be read and analyzed in class.

FEAR AND THE STUDENT CHEATER*

by Charles R. Tittle and Alan R. Rowe†

Classroom cheating has plagued universities for generations and solutions to the problem are constantly being sought. The suggestion that the solution lies in a moral appeal to a "community of scholars" rather than in threats of coercion is heard more and more. It is argued that adherence to academic ethics can best be achieved through mutual trust and recognition of the fundamental moral foundations of scholarly norms. This position has been bolstered by a belief that punishment has little influence on behavior anyway, and it is consistent with neo-Enlightenment philosophies that are popular among students.

The value of a moral appeal to encourage honesty has been countered by a body of empirical evidence concerning punishment. Recent investigations by Charles Tittle and Charles Logan (see *Law and Society Review*, 7, Spring 1973) suggest that negative sanctions can be effective in deterring deviance. It still remains to be established, however, just how much impact sanctions have on behavior under various kinds of conditions, and how important threatened punishment is relative to other variables, such as moral appeals.

A Test of Effectiveness

We set out to determine the relative effectiveness of simple trust, a punishment threat, and a moral appeal in achieving classroom honesty. Three sociology classes served as subjects. One class was designated as a control group while the other two (which were taught in the quarter immediately following) were the test groups. Questionnaire data to enable comparison of the classes and analysis of the impact of the threat on various categories of individuals were gathered at the beginning of each of the courses. Students were told that the information requested

---

* This article is reprinted with permission from *Change Magazine*, April, 1974. A more detailed report of this experiment appears in *Social Problems*, Vol. 20, No. 4, Spring 1973.

† Charles R. Tittle is a professor of sociology and Alan R. Rowe is an assistant professor of sociology at Florida Atlantic University.

would make it possible to tailor the level of presentation to the needs and capabilities of the students.

The experiment consisted of organizing each of the three classes around a series of eight weekly quizzes worth 10 points each. The quizzes were administered during the last 15 minutes of a class period and were collected as the students left. After class, the scores were recorded by the instructors (who were also the experimenters) but no marks were made on the students' papers. At the beginning of the next class period the quizzes were returned and the students were told to calculate their own grades. A difference between the score obtained by the instructor and the score which the student calculated was used as an indication of cheating. To minimize moral and ethical dilemmas for the experimenters, the students' self-assigned scores were the ones used in the calculation of course grades, and the experimenters agreed to keep the identity of cheaters secret, even from each other.

In the control group no mention was made of cheating, the necessity for honesty, or of the possibility of being caught or punished. It was simply explained that taking the quizzes at the end of the period was less distracting to the lecture of that day and that this procedure permitted individuals flexibility in the time they devoted to a quiz. The two test groups were exposed to this same procedure for three quizzes; but before they graded their fourth quiz, these classes were reminded that they were being trusted to grade their quizzes honestly and that they had a moral obligation to be accurate. No other mention was made of cheating until the seventh quiz was returned. At that point the students were told that there had been complaints about cheating, so it was necessary for the instructor to spot check some of the quizzes for accuracy. When the eighth quiz was returned, the students were told that the spot check had revealed a case of cheating and that the person was to be penalized (in fact, nobody was penalized).

Only students who took and graded at least one quiz of the first six and of the last two were considered for the analysis, and then only if they had some reason to consider cheating on one of the two series (that is, if they had given an incorrect answer on a quiz). When other series of quizzes were compared, the same method of exclusion was applied. As a result, the number of quizzes considered in a given series varies depending upon the comparisons made.

Cheating and Trust

By comparing the percentages of cheating opportunities taken by students at various points in the test classes with the figures for the control group, it is possible to ascertain the effects of the three approaches for controlling cheating. Table 1 reports the data. They show, first, that the incidence of cheating under conditions of "trust" was astonishingly high. Students who, with no specific moral appeal, were trusted to be honest actually cheated about one third of the time (quizzes 1–6 for groups A and B, and quizzes 1-8 for the control group). Clearly, simple trust is not

TABLE 1 MEAN PERCENT OF CHEATING OPPORTUNITIES UTILIZED

Test Group	With No Appeal or Threat (Quiz 1–3)	After Moral Appeal (Quiz 4–6)	After Spot-Check Threat (Quiz 7)	After Sanction Threat (Quiz 8)
A ($N = 30$)	31% ($N = 30$)	41% ($N = 39$)	13% ($N = 30$)	11% ($N = 26$)
	34% ($N = 30$)		12% ($N = 30$)	
B (N = 51)	41% ($N = 47$)	43% ($N = 47$)	32% ($N = 46$)	22% ($N = 29$)
	42% ($N = 51$)		31% ($N = 51$)	
Control ($N = 26$)	27% ($N = 26$)	33% ($N = 24$)	24% ($N = 26$)	28% ($N = 12$)
	30% ($N = 26$)		24% ($N = 26$)	

enough. The data show further that emphasizing the moral principle involved in grading the quizzes was also ineffectual. A moral appeal had no effect whatsoever in reducing the incidence of cheating. Consider, for example, test group A. Prior to the moral appeal, the students took 31 percent of all opportunities to cheat, but after the moral appeal this percentage actually rose to 41 percent.

The data do demonstrate that the threat of being caught and punished had a substantial effect on cheating. In test group A, the cheating level of 34 percent was reduced to 12 percent by the threat of a spot check and the ultimate announcement of a penalty for a "cheater." A similar but less marked pattern characterized test group B. Thus, from the data presented in Table 1, it seems clear that fear of sanction has a greater influence on classroom honesty than does simple trust or a moral appeal.

There is the possibility that the test groups differed from the control group in some way that may have accounted for the effect of the punishment threat and the failure of the moral appeal. To evaluate this, the classes were compared in terms of age, sex, reason for enrolling, expected grade, motivation for earning a high grade, discrepancy between the expected grade at the beginning of the quarter and the actual grade earned, and the major area of study. No significant differences were found between the control and test groups that could have accounted for the results. Moreover, there is some "informal" indication that the slight decline in cheating in the control group (from 30 percent to 24 percent) after the sixth quiz was itself attributable to increased fear of punishment near the end of the quarter. A student in another class inquired of one of the authors, point blank, if he were going to check some of the quiz papers. She had heard rumors that students in that class were growing fearful that the "grade-your-own" policy had resulted in so much cheating that the instructor would surely do something before the end of the quarter.

The Certainty of Sanction

The two test groups (A and B) apparently did differ in one important aspect—the certainty of being sanctioned. Not only was group B more than twice as large as group A so that a spot check in group A actually implied a greater probability of being caught, but the instructor for group B had a reputation among students as lovable and understanding. Thus, fear of punishment was probably more credible in group A than in group B even before the declaration prior to quiz eight that a cheater had been discovered and penalized.

Not only did the sanction threat have more effect than the moral appeal, but it affected some individuals more than others. Females cheated less initially and were far more responsive to the threat of being caught and punished. The females in our study fit the profile developed through years of social psychological research: they were more conforming, more obedient to authority, and less willing to take chances than males. Presumably women in our society have a greater consciousness of status and sensitivity to reputation, both of which would intensify fear of exposure for dishonest behavior.

Cheating and Grades

We found that cheating was related to a student's grade expectations. Those who were doing worst in the course and who were experiencing the greatest discrepancy between the grade they said they expected to earn and the grade they actually earned cheated more than others and were less responsive to the sanction threat. In short, those who were most in need of points were willing to take greater risks. This is consistent with the theory that the greater the utility of an act, the greater the potential punishment required to deter it. And perhaps it shows the futility of a moral appeal in a social context where all individuals are not successful.

All in all, the findings do not bode well for moralistic arguments. To be sure, moral principles were no doubt operating to an extent: there were some students who cheated only a little even when they needed a certain grade very badly, and a few students actually gave themselves lower grades than they earned. Nevertheless, these cases were exceptional; most students cheated until they were afraid of the consequences.

While it undoubtedly would be a mistake to draw sweeping conclusions from the results of the experiment, the data suggest the applicability to the college teacher (at least in his role as evaluator) of Machiavelli's cogent observation: since it is difficult to be both loved and feared, "it is much safer to be feared than loved, if one must choose."

ASSIGNMENT

Here is one question you might try to answer using topics covered in this textbook. You can make up others like it.

1 Was there a significant difference between the mean percentage of cheating opportunities utilized in group A and the control group before any appeal or threat was made? (Assume $s_A = .08$ and $s_B = .09$ since not enough data are given in the article.)
2 Discuss the article in class, especially the way the experiment was set up. Do you think such an experiment would get similar results on your campus?

BEAUTY DOES AS BEAUTY IS: HOW LOOKS INFLUENCE LIKING*

Beauty may be only skin deep, but its effects run considerably deeper. Ellen Berscheid and Elaine Walster's article, "Beauty and the Best" [*Psychology Today,* March 1972] is one of many studies confirming that men and women generally like attractive people more, expect them to do better work and rate their work more highly after it's done.

Psychologists David Landy and Harold Sigall decided to see which was cause, and which was effect, in the relationship between a woman's physical attractiveness and the way others rate her work. Perhaps it is not a matter of observer bias; perhaps attractive people actually work better, because of genetic or environmental advantages.

Landy and Sigall prepared two essays on television's role in society: one well written, the other disorganized and simplistic in its ideas. They made 30 copies of each essay and attached pictures, supposedly of the author, to 20 of them: an attractive woman in 10 cases, an unattractive one in the other 10. Ten essays in each group had no picture.

The experimenters gave the essays to 60 male undergraduates, who rated them for creativity, ideas, style and general quality, and who also rated the author's intelligence, sensitivity, talent and ability.

The attractive writers swept the ratings race across the board, and they most surpassed the unattractive writers when the poor essay was judged. Landy and Sigall conclude that "if you are ugly you are not discriminated against a great deal so long as your performance is impressive. However, should performance be below par, attractiveness matters: you may be able to get away with inferior work if you are beautiful . . . Perhaps this expectancy [that attractive people will do well] leads evaluators to give physically attractive performers the benefit of the doubt when performance is substandard or of ambiguous quality."

---

* This article is reprinted with permission from Newsline in PSYCHOLOGY TODAY Magazine, August 1974.

The researchers admit that their interpretation is still speculative, and based only on male responses to female attractiveness. But, they note, their results are consistent with other findings that show discrimination based on "irrelevant" characteristics, and "that both men and women are rather susceptible to having their judgments influenced by the physical attractiveness of others."—Jack Horn

Here is some further data taken from the original report published in the *Journal of Personality and Social Psychology* Vol. 29, No. 3 where it was analyzed by methods not covered in this book. But we can ask some questions which you can answer using topics covered in this book.

Table 2 shows the ratings given to the essays for "General Quality of Essay."

TABLE 2

Quality	Picture of Attractive Woman	No Picture	Picture of Unattractive Woman
Good Essay	$m = 6.7$	$m = 6.6$	$m = 5.9$
	$s = 1.6$	$s = 1.4$	$s = 1.6$
Poor Essay	$m = 5.2$	$m = 4.7$	$m = 2.7$
	$s = 1.6$	$s = 2.0$	$s = 1.3$

m is the mean score given by 10 male readers. s is the standard deviation. Each reader gave the essay a score from 1 to 9.

ASSIGNMENT

3 Is there a significant difference at $\alpha = .05$ between the ratings the attractive and unattractive women got on the good essays? On the poor essays?

4 Discuss the experiment. Do you believe you would get similar results on your campus? Of what value is this experiment?

JURY VERDICTS LINKED TO NEWS REPORTS*

Two Columbia University social scientists said yesterday that a three-year study had yielded a preliminary finding that prejudicial news coverage of pretrial evidence adversely affected jury verdicts.

Dr. Alice Padawer-Singer and Dr. Allen H. Barton of Columbia's Bureau of Applied Social Research, found evidence that jurors exposed to prejudicial news stories were as much as 66 percent more likely to find defendants guilty than jurors who read objective news reports.

Their findings uphold the contentions of such highly publicized persons as Spiro T. [Agnew] and John N. Mitchell, who have said that they could not be tried fairly because of prejudicial news articles printed in advance of their trials.

In experiments recreating an actual case and courtroom conditions, the researchers also found that jurors who were not screened for impartiality through customary "voir dire" examinations were more likely to return guilty verdicts.

The study, to be published in December, has led to a second grant, this one for $571,500 from the National Science Foundation, to examine the merits of six-person and twelve-person panels and to compare the advantages of unanimous and non-unanimous verdicts.

The original study was financed by the National Science Foundation, the Columbia Broadcasting System, The New York Times, the Russell Sage Foundation, the Walter E. Meyer Research Institute of Law and the Lawrence E. Wien Foundation.

Selecting at random 400 people from actual jury pools in the state supreme courts in Brooklyn and Nassau County, the scientists, with the aid of two lawyers, re-enacted the jury-selection process and an actual Washington murder trial that had taken place the previous year.

Before the trial, half the jurors read prejudicial news stories concerning the case and the others were given objective clippings.

In one set of experiments, the Columbia researchers found that 33 of 60 jurors who read "straight" articles thought the defendant was guilty, while 47 of 60 who were exposed to prejudicial publicity voted guilty—a difference of 23 percent.

In another group of tests, the figures were seven jurors for guilty of 60 who read straight articles and 47 for guilty of 60 who read prejudicial stories—66 percent more.

The scientists found, however, that jurors who had been screened for prejudice before the trial were less likely to be influenced by prejudicial news article, even if they were exposed to such articles.

"Strenuous efforts should be made throughout the nation's courts to educate jurors away from their personal biases as well as from the effects of prejudicial news stores," Dr. Padawer-Singer said.

ASSIGNMENT

5 Is the 23 percent difference referred to in the first set of experiments significant at $\alpha = .05$?

ANSWERS TO SELECTED ASSIGNMENTS

CHAPTER 1

1-3 More

CHAPTER 3

3-1 False. Two outcomes does not imply that they are equally likely.

3-3 (a)

	1	2	3	4	5	6	7	8	9
1st rod	1	1	1	2	2	2	3	3	3
2d rod	1	2	3	1	2	3	1	2	3

(b) $X = 4$ in 3 cases: (1,3), (2,2), (3,1); therefore $P(X = 4) = 3/9 = .33$
(c) $X > 4$ in 3 cases: (2,3), (3,2), (3,3); therefore $P(X = 4) = 3/9 = .33$

3-5 (a) $\pi = P$ (a biology major is female) (b) An estimate of π (c) We do not know (d) $40/60 = .667$ (e) $20/60 = .333$

3-7 (a) See Table A3-7. (b) $P(X = 4) = 1/16 = .0625$ (c) No (d) .0625

TABLE A3-7

	Guess on No. 1	Guess on No. 2	Guess on No. 3	Guess on No. 4
1	g	g	g	g
2	g	g	g	b
3	g	g	b	g
4	g	g	b	b
5	g	b	g	g
6	g	b	g	b
7	g	b	b	g
8	g	b	b	b
9	b	g	g	g
10	b	g	g	b
11	b	g	b	g
12	b	g	b	b
13	b	b	g	g
14	b	b	g	b
15	b	b	b	g
16	b	b	b	b

3-9 These statements were probably arrived at by taking a sample from the appropriate population and then computing the percent of the population that fits the category described.

3-11 (a) 1 (b) 0

CHAPTER 4

4-1

	(a)	(b)	(c)	(d)	(e)	(f)	(g)	(h)
df	8	5	21	3	6	4	7	21
X_c^2	20.09	15.09	32.67	7.82	16.81	9.49	14.07	38.93

4-3 (a) B, C (b) A (c) B

4-7

	1	2	3	4	5	6
O	7	8	15	10	11	9
E	10	10	10	10	10	10

H_0: The die is fair.

df = 5

All E's > 5 $X^2 = \Sigma(O^2/E) - n = 4.9 + 6.4 + 22.5 + 10 + 12.1 + 8.1 - 60 = 4$
For $\alpha = .05$ $X_c^2 = 11.07$ Since $X^2 < X_c^2$ we do not reject H_0. We do not have evidence to prove the die is not fair.

4-9 (a)

	R	L	A
O	80	17	3

$n = 100$

(b)

	R	L	A
E	74	20	6

$n = 100$

All E's > 5
(c) $3 - 1 = 2$ (d) The theory is correct. (e) $X_c^2 = 5.99$ (f) If H_0 is true, 95 percent of experimental X^2's should be less than 5.99. (g) $X^2 = 6{,}400/74 + 289/20 + 9/6 - 100 = 2.44$. No. 2.44 is less than 5.99

4-11 Table of expected values if H_0 is true:

	Black	Spanish	Jewish	White Prot.	White Cath.	
Bean	182.14	109.29	72.86	72.86	72.85	510
Barildo	317.86	190.71	127.14	127.14	127.15	890
	500	300	200	200	200	$n = 1{,}400$

All E's > 5

Cell	O^2	O^2/E
1	10,000	54.90
2	2,500	22.87
3	10,000	137.25
4	12,100	166.07
5	22,500	308.85
6	160,000	503.37
7	62,500	327.72
8	10,000	78.65
9	8,100	63.71
10	2,500	19.66

df = 4

$\Sigma(O^2/E) = 1{,}683.05$

$X^2 = 283.05$

$X_c^2 = 13.28$

$X^2 > X_c^2$ Therefore, we reject H_0. We have proof that there is dependence between ethnic group and voting preference.

4-13 Expected values if H_0 is true:

	F	A	NO	
M	21.84	21.84	8.32	52
S	20.16	20.16	7.68	48
	42	42	16	100

df = 2

$\alpha = .01$

$X_c^2 = 9.21$

All E's > 5

Cell	1	2	3	4	5	6
O^2/E	10.30	26.37	20.31	36.16	16.07	1.17

$X^2 = \Sigma(O^2/E) - n = 110.38 - 100 = 10.38$ $X^2 > X_c^2$ implies there is a relation between marital status and attitude toward women's liberation.

4-15 Expected values if H_0 is true:

	Before 12*N*	12*N*–3*P*	3*P*–5*P*	5*P*–7*P*	After 7*P*	
IA	9.4	12	11.2	10.4	7	50
SA	9.4	12	11.2	10.4	7	50
CA	9.4	12	11.2	10.4	7	50
CC	9.4	12	11.2	10.4	7	50
HA	9.4	12	11.2	10.4	7	50
	47	60	56	52	35	250

All E's > 5

Cell	1	2	3	4	5	6	7	8
O²/E	42.55	8.33	22.86	1.54	0	1.70	21.33	7.23

Cell	9	10	11	12	13	14	15	16	17
O²/E	18.85	7	8.62	18.75	20.09	4.71	2.29	8.62	4.08

Cell	18	19	20	21	22	23	24	25
O²/E	2.23	34.71	14.29	2.66	12	10.8	6.15	28

df = 16 $\quad \alpha = .01 \quad X_c^2 = 32.00 \quad X^2 = \Sigma(O^2/E) - n = 309.39 - 250 = 59.39$

$X^2 > X_c^2$ Therefore, we have evidence that there is a relation between accident type and time of occurrence.

4-17 Table of expected values:

	CD	NCD	
Fr.	18.92	11.08	30
So.	29.65	17.35	47
Jr.	14.51	8.49	23
Sr.	18.92	11.08	30
Total	82	48	130

df = 3
$\alpha = .01$
$X_c^2 = 11.34$

All E's > 5 $\quad X^2 = \Sigma(O^2/E) - n = (25.58 + 5.78 + 43.71 + 6.97 + 4.41 + 26.50 + 13.53 + 17.69) - 130 = 14.17$ Since $X^2 > X_c^2$ we have evidence that college class and decision about careers are related.

4-19 Table of expected values:

	Seldom	Sometimes	Very Often	
Over 5 ft 7 in.	7.25	11	6.75	25
5 ft 2 in. to 5 ft 7 in.	15.66	23.76	14.58	54
Under 5 ft 2 in.	6.09	9.24	5.67	21
	29	44	27	100

df = 4
$\alpha = .05$
$X_c^2 = 9.49$

All E's > 5

Cell	1	2	3	4	5	6	7
O^2/E	13.79	11	2.37	12.52	28.45	13.44	4.11

Cell	8	9
O^2/E	5.30	14.29

$X^2 = \Sigma(O^2/E) - n = 105.27 - 100 = 5.27$ X^2 is not greater than X_c^2 so we do not have sufficient evidence to claim that there is a relationship between height and wearing heels.

4-21

	Ed	S.S.	Hum.	B.S.	N.S.	A.H.	Other	
M	6.75	8.34	8.73	9.13	5.56	6.75	6.74	52
F	10.25	12.66	13.27	13.87	8.44	10.25	10.26	79
	17	21	22	23	14	17	17	131

All E's > 5 df = 6 $\alpha = .01$ $X_c^2 = 16.81$

Cell	1	2	5	4	5	6	7	8	9
O^2/E	2.37	11.99	0.46	28.04	14.57	2.37	7.27	16.49	9.56

Cell	10	11	12	13	14
O^2/E	30.14	3.53	2.96	16.49	9.75

$X^2 = 155.99 - 131 = 24.99$ $X^2 > X_c^2$. There is evidence of a relationship between sex and major field at Podark C.C.

4-23

	S	G	P	
S	33.33	33.33	33.34	100
G	33.33	33.33	33.34	100
P	33.34	33.34	33.32	100
	100	100	100	300

df = 4
$\alpha = .01$
$X_c^2 = 13.28$

All E's > 5

Cell	1	2	3	4	5	6	7	8	9
O^2/E	75	18.75	18.75	18.75	75	18.75	18.75	18.75	75

$X^2 = 337.50 - 300 = 37.50$ $X^2 > X_c^2$ We have evidence of a relationship between the eye type of mother robots and child robots.

4-25

	1	2	3	4	
1	44.55	41.36	41.36	47.73	175
2	49.64	46.09	46.09	53.18	195
3	45.81	42.55	42.55	49.09	180
	140	130	130	150	550

$df = 6$
$\alpha = .05$
$X_c^2 = 12.59$

All E's > 5

Cell	1	2	3	4	5	6	7	8
O^2/E	35.91	38.68	48.96	52.38	50.36	43.94	49.99	50.85

Cell	9	10	11	12
O^2/E	54.57	47.59	32.17	46.93

$X^2 = 552.33 - 550 = 2.33$ $X^2 \not> X_c^2$ We do not have sufficient evidence to say that there is a relationship between starting position and finishing place for the first 4 starting positions.

4-27 He could gather more observations.

4-37 Adjusted observed values

	Mangos	Guavas	
Left-handed	28 + .5	42 − .5	70
Right-handed	19 − .5	11 + .5	30
	47	53	100

Expected values

	Mangos	Guavas
Left-handed	32.9	37.1
Right-handed	14.1	15.9

$X^2 = 103.70 - 100 = 3.70 < X_c^2 = 3.84$. Fail to reject H_0.

CHAPTER 5

5-1 True

5-3 Company A. The smaller standard deviation means their pieces of wood have less variability. They are more consistent.

5-5 (a) $\mu_W \pm 2.58\ \sigma_W$ $180 \pm 2.58(10)$ 180 ± 25.8 154.2 lb; 205.8 lb (b) $z_{173} = (173 - \mu)/\sigma = (173 - 180)/10 = -7/10 = -.7$ (c) $W = \mu_W + z(\sigma_W) = 180 + (2.1)(10) = 180 + 21 = 201$ lb

5-7 (a) To the left of $z = 2$ we have .977. To the left of $z = -2$ we have .023. Therefore, between 2 and −2 we have $.977 - .023 = 9.954 = 95.4\%$ (b) To the left of $z = 1$ is .84. Therefore, to the right of $z = 1$ is $1.00 - .84 = .16 = 16\%$ (c) Left of $z = 1.96$ is .975. Left of $z = 1$ is .84. Therefore, between them is $.135 = 13.5\%$ (d) Proportion of area left of $z = +3$ is .999. Left of $z = -3$ is .001. Therefore, between −3 and +3 is $.998 = 99.8\%$ (e) Left of $z = -2$ is 2.3% (f) From part a, 95.4% is within 2 standard deviations. Therefore, $100\% - 95.4\% = 4.6\%$ is more than 2 standard deviations away from the mean.

5-9 (a) $z_{49} - 2$; 2; 3% (b) $z_{34} = -3$; 99.9% (c) 68% (d) 35.26 (e) 37.12 and 48.88

5-11 Cannot be answered since shape of distribution is unknown.

5-13 (a) $1/6 = .167$ (b) P (a three is not rolled) (c) $5/6 = .833$ (d) $\mu_p = \pi = .167$ (e) $\sqrt{(.167)(.833)/100} = .037$ (f) $\mu_p \pm 1.96\ \sigma_p$ $.167 \pm (1.96)(.037)$ $.167 \pm .073$ .094, .240

5-15 (a) Chi-square (b) binomial (c) neither (d) binomial (e) chi-square

5-17 Population: All possible tosses of this coin. Let $\pi = P(\text{heads})$ $H_0 : \pi = .5$ (coin is fair) $H_a : \pi \neq .5$ $n = 100$ $n\pi = 50 > 5$ $n(1 - \pi) = 50 > 5$ $p = 37/100 = .37$ $\mu_p = \pi = .50$ $\sigma_p = \sqrt{(.50)(.50)/100} = .05$ $\alpha = .01$ Therefore, $z_c = \pm 2.58$. $p_c = \mu_p + (z_c)\sigma_p = .50 + (\pm 2.58)(.05) = .50 \pm .129 = .371, .629$ p is not between .371 and .629. Therefore, we can claim evidence that the coin is not fair. Note, however, the outcome

$p = .37$ is very close to one of the critical values. If possible it would be a good idea to toss the coin more times to see if the patterns are consistent.

5-19 Population: College students $H_0: \pi = .50$ (50% are moderate or conservative) $H_a: \pi \neq .50$ $n = 300$ $n\pi = 150 > 5$ $n(1-\pi) = 150 > 5$ $\mu_p = .50$ $\sigma_p = \sqrt{(.5)(.5)/300} = .029$ $\alpha = .05$ $z_c = \pm 1.96$ $p_c = .50 + (\pm 1.69)(.029) = .50 \pm .057 = .443, .557$ $p = 140/300 = .467$ There is not sufficient evidence to suggest that π is different from .50.

5-21 Population: People who contract this disease $H_0: \pi = .50$ (50% of them had measles before age 2) $n = 90$ $n\pi = 45 > 5$ $n(1-\pi) = 45 > 5$ $\mu_p = .50$ $\sigma_p = \sqrt{(.50)(.50)/90} = .053$ $p_c = .50 \pm (1.96)(.053) = .50 \pm .104 = .396, .604$ $p = 40/90 = .44$ This tends to confirm the theory since it gives us no evidence to reject $\pi = .50$.

5-23 Population: Students on your campus $H_0: \pi = .60$ (60% disapprove of the law) $n = 100$ $n\pi = 60 > 5$ $n(1-\pi) = 40 > 5$ $\mu_p = .60$ $\sigma_p = \sqrt{(.60)(.40)/100} = .049$ $\alpha = .05$ $p_c = .60 \pm 1.96(.049) = .60 \pm .096 = .504, .696$ $p = 44/100 = .44$ Indicates the percent on campus is significantly lower than the percent in the general U.S. population

5-25 Population: All possible draws from a shuffled deck $H_0: \pi = .25$ (25% of draws yield diamonds) $n = 100$ $n\pi = 25 > 5$ $n(1-\pi) = 75 > 5$ $\mu_p = .25$ $\sigma_p = \sqrt{(.25)(.75)/100} = .043$ $\alpha = .01$ $p_c = .25 \pm (2.58)(.043) = .25 \pm .111 = .139, .361$ $p = 21/100 = .21$ We would have conclusive evidence that diamonds were not drawn 25 percent of the time if we got in a sample of 100 draws either less than 14 or more than 36 diamonds. Since this did not happen, we do not have compelling evidence that diamonds are coming up less than expected.

5-27 Population: Female college students at Meg's school H_0: $\pi = .65$ (65% feel they would "lose femininity") $n = 80$ $n\pi = 52 > 5$ $n(1-\pi) = 28 > 5$ $\mu_p = .65$ $\pi_p = \sqrt{(.65)(.35)/80} = .053$ $\alpha = .05$ $p_c = .65 \pm 1.96(.053) = .65 + .10 = .55, .75$ $p = 42/89 = .47$ This is significantly lower than the results at the Midwestern University. The women at Meg's school evidently do not agree.

5-29 Population: All people living in the United States $H_0: \pi = .01$ (1% died this year) $n = 192{,}000$ $n\pi = 1920 > 5$ $n(1-\pi) = 190{,}080 > 5$ $\mu_p = .01$ $\sigma_p = \sqrt{(.01)(.99)/192{,}000} = .00023$ $\alpha = .05$ $p_c = .01 \pm 1.96(.00023) = .01 \pm .00045 = .00955, .01045$ $p = 1{,}800/192{,}000 = .009375$ This is significantly lower than $\pi = .01$. Evidently the percent of people dying has changed. It is now lower than it was in 1950. Whether or not the new figure is *enough* lower to be of any *practical* concern is a separate question.

5-31 (a) $n\pi = 7.5$ $n(1 = \pi) = 37.5$ $\mu_p = \pi = 1/6 = .17$ $\sigma_p = \sqrt{(1/6)(5/6)/45} = \sqrt{1/324} = 1/18 = .056$ $p_c = 1/6 \pm 1.96(1/18) = .17 \pm 1.96(.056) = .17 \pm .11 = .06, .28$ (b) One (c) B

CHAPTER 6

6-1 (a) Binomial one sample (b) binomial two sample (c) other (d) other (e) chi-square

6-5 Results of one experiment we did:

	Head	Tail
Spin	10	15
Toss	13	12

All outcomes > 5 $p_1 = 10/25 = .40$ $p_2 = 13/25 = .52$ $p = (10 + 13)/(25 + 25) = 23/50 = .46$ $s_{dp} = \sqrt{\frac{(.46)(.54)}{25} + \frac{(.46)(.54)}{25}} = \sqrt{.019972} = \sqrt{.0200} = .1414$

$dp_c = 0 \pm 1.96(.1414) = \pm.28$ $dp = .40 - .52 = -.12$ Since $-.28 < -.12 < +.28$ we did not have enough evidence to show that the percent of heads is different for these two techniques.

6-7 40, 20, 60, 20 > 5 $n_1 = 60$ $n_2 = 80$ $\alpha = .05$ $p_1 = 40/60$ $p_2 = 60/80$

$p = (40 + 60)/(60 + 80) = .714$ $s_{dp} = \sqrt{\frac{(.714)(.286)}{60} + \frac{(.714)(.286)}{80}} = .077$

$dp_c = 0 \pm 1.96(.077) = \pm.15$ $dp = .67 - .75 = -.08$ This is not a big enough difference to claim that the potion works differently on right- and left-handed people.

6-9 $\pi_1 = P$(a nematode survives on solid heme) $\pi_2 = P$(a nematode survives on liquid heme) 400, 100, 300, 100 > 5 $p_1 = 400/500 = .80$ $p_2 = 300/400 = .75$

$p = (400 + 300)/(500 + 400) = .778$ $H_0: \pi_1 = \pi_2$ $s_{dp} = \sqrt{\frac{(.778)(.222)}{500} + \frac{(.778)(.222)}{400}} = .0279$

$dp_c = 0 \pm 2.58(.0279) = \pm.072$ $dp = .80 - .85 = .05$ Fail to reject H_0. This does not support the theory that the heme acts differently in the liquid and solid states.

6-11 30, 50, 40, 10 > 5 $n_1 = 80$ $n_2 = 50$ $\alpha = .05$ $p_1 = 30/80 = .375$ $p_2 = 40/50 = .80$

$p = (30 + 40)/(80 + 50) = .538$ $s_{dp} = \sqrt{\frac{(.538)(.462)}{80} + \frac{(.538)(.462)}{50}} = .090$

$dp_c = 0 + (1.96)(.090) = \pm.176$ $dp = .375 - .80 = -.425$ This is a large enough difference for us to claim that these two groups (parents and nonparents) contain different percents of people favoring more lenient laws with regard to marijuana possession.

6-13 15, 85, 20, 80 > 5 $n_1 = 100$ $n_2 = 100$ $\alpha = .05$ $p_1 = 15/100 = .15$ $p_2 = 20/100 = .20$

$p = (15 + 20)/(100 + 100) = .175$ $s_{dp} = \sqrt{\frac{(.175)(.825)}{100} + \frac{(.175)(.825)}{100}} = .0537$

$dp_c = 0 \pm 1.96(.0537) = \pm.105$ $dp = .15 - .20 = -.05$ This is not a big enough difference to prove that type B is more effective than type A.

6-15 190, 20, 30, 160 > 5 $n_1 = 210$ $n_2 = 190$ $\alpha = .01$ $p_1 = 190/210 = .905$

$p_2 = 30/190 = .158$ $p = (190 + 30)/(210 + 190) = .55$ $s_{dp} = \sqrt{\frac{(.55)(.45)}{210} + \frac{(.55)(.45)}{190}} = .050$

$dp_c = 0 \pm 2.58(.050) = \pm.129$ $dp = .905 - .158 = .747$ This difference is much greater then the critical difference. Therefore the percent of type A moths attracted by the first chemical ratio is evidently quite different from the percent of type A moths attracted by the second chemical ratio. Population 1 is "all male moths attracted by the first chemical ratio." π_1 is the percent of type A moths in population 1.

6-23 $H_0: \pi = .10$ $H_a: \pi \neq .10$ $n\pi = 40 > 5$, $n(1 - \pi) = 360 > 5$ $\sigma = \sqrt{\frac{(.1)(.9)}{400}} = .015$

$p_c = .10 \pm 2.58(.015) = .10 \pm .04 = .06$ and $.14$ $p = 20/400 = .05$ We fail to reject H_0; 10% may be correct.

CHAPTER 7

7-9 $H_0: \mu = 100$ (the British average is also 100) $\sigma_{pop} = 10$ (We assume that the British scores also have $\sigma_{pop} = 10$) $n = 64$ $\alpha = .05$ $\sigma_m = 10/\sqrt{64} = 1.25$ $m_c = 100 \pm 1.96(1.25) = 100 \pm 2.45 = 97.55, 102.45$ $m = 105$ Evidently the British score higher on average than the Americans do.

7-11 (a) $m = 204/6 = 34$ years (b) $m = 238/7 = 34$ years (c) In B the average is more representative. When the numbers are close to the average, the average is more representative.

7-13

X	40	32	60	80	45	40	80	71	20	16	42
X^2	1,600	1,024	3,600	6,400	2,025	1,600	6,400	5,041	400	256	1,764

X	37	80	75	69	80
X^2	1,369	6,400	5,625	4,761	6,400

$\Sigma X = 867 \quad (\Sigma X)^2 = 751{,}689 \quad n = 16 \quad \Sigma X^2 = 54{,}665 \quad m = 867/16 = 54.19$

$$s_{\text{pop}} = \sqrt{\frac{54{,}665 - 751{,}689/16}{15}} = \sqrt{\frac{54{,}665 - 46{,}980.56}{15}} = \sqrt{7{,}684.44/15} = \sqrt{512.30} = 22.63$$

$s_m = 5.66$

7-15 In Assignment 7-12, since $n > 30$, we can think of m as a member of the normal distribution of all possible m's for samples of size 36 from the population of that exercise.

7-17 $\alpha = .01 \quad z_c = \pm 2.58 \quad n = 400 \quad H_0\colon \mu = 5$ ft 10 in. $= 70$ in. (the average is 70 in.) $s_{\text{pop}} = 4.2$ in. $\quad s_m = 4.2/20 = .21$ in. $\quad m_c = 70 \pm 2.58(.21) = 70 \pm .542 = 69.46, 70.54$ $m = 70.8$ in. This value of m is large enough to allow us to claim that the average height is not 5 ft 10 in., but is more than that. The probability that the average height is 5 ft 10 in. is less than 1%.

7-19 $\alpha = .01 \quad z_c = \pm 2.58 \quad n = 100 \quad H_0\colon \mu = 2.91$ (The mean cumulative average is 2.91.) $s_{\text{pop}} = 3.0 \quad s_m = 3.0/10 = .3 \quad m_c = 2.91 \pm 1.96(.3) = 2.91 \pm .57 = 2.34, 3.48 \quad m = 2.86$ These figures do not show any significant difference between the grades in the sorority and the grades of female students in general.

7-21 $\alpha = .05 \quad z = \pm 1.96 \quad n = 50 \quad H_0 : \mu = 5.5 \quad s_{\text{pop}} = 1.8 \quad s_m = 1.8/7.07 = .255$ $m_c = 5.5 \pm (1.96)(.255) = 5.5 \pm .500 = 5, 6 \quad m = 4.2$ This indicates that either (1) the people in Suffolk County watch as much TV as anyone else, but that our sample was exceptional, or (2) the people in Suffolk County tend to watch TV less hours per day than in other parts of the county.

7-23 $\alpha = .05 \quad z_c = \pm 1.96 \quad n = 40 \quad H_0 : \mu = 80 \quad s_{\text{pop}} = \sqrt{\dfrac{269{,}237 - (3{,}227)^2/40}{39}} =$

$\sqrt{\dfrac{269{,}237 - 104{,}135.29/40}{39}} = 15.105 \quad s_m = 15.105/6.325 = 2.388 \quad m_c = 80 \pm 1.96(2.388) =$

$80 \pm 4.68 = 75.3, 84.7 \quad m = 3{,}227/40 = 80.675$ This indicates the average pulse rate in this hospital is about right. However, this may not be very important or useful information. One might guess that patients in a hospital would tend to have either unusually high or unusually low pulse rates. If that is true, then the average at the hospital would be the normal average, but the variance at the hospital would be much larger than the normal variance. Tests for abnormal variance can be found in the references at the end of the text.

7-25 $t_c = \pm 2.92$ **7-27** $t_c = \pm 3.17$

7-29 (a) Binomial one sample (b) sample mean, large sample (c) chi-square (d) other (e) binomial two sample (f) binomial one sample (g) other (h) sample mean, small sample

7-31 $\alpha = .01 \quad n = 20 \quad \text{df} = 19 \quad t_c = \pm 2.86 \quad H_0 : \mu = 32 \quad s_{\text{pop}} = 1.4 \quad s_m = 1.4/4.472 = .313$ $m_c = 32 \pm (2.86)(.313) = 32 \pm .895 = 31.105, 32.895 \quad m = 30.8$ This evidence tends to deny the claim. It appears that the average age is younger than 32.

7-33 $\alpha = .01 \quad n = 9 \quad \text{df} = 8 \quad t_c = \pm 3.36 \quad H_0 : \mu = 80 \quad s_{\text{pop}} = 12 \quad s_m = 12/3 = 4$ $m_c = 80 \pm 3.36(4) = 80 \pm 13.44 = 66.56, 93.44 \quad m = 70$ This sample data supports the report. $m = 70$ is not unusual for a small random sample from a population with $\mu = 80$.

7-35 $\alpha = .05$ $n = 12$ $\text{df} = 11$ $t_c = \pm 2.20$ $H_0: \mu = 3225$ $s_{pop} = 500$ $s_m = 500/3.464 = 144.34$ $m_c = 3225 \pm 2.20(144.34) = 3225 \pm 317.55 = 2907.45, 3542.55$ $m = 3{,}500$ This is within the critical values. We have no strong evidence that the magazine is wrong.

7-37 $\alpha = 01$ $n = 50 > 30$ $z_c = \pm 2.58$ $H_0: \mu = 100$ $s_{pop} = 14$ $s_m = 14/7.07 = 1.98$ $m_c = 100 \pm 2.58(1.98) = 100 \pm 5.11 = 94.89, 105.11$ $m = 105$ It is possible based on these figures that porpoises have a higher average PIQ than 100, but the sample mean falls just within the critical value. We cannot say we have proved at the .01 significance level that they have higher PIQs. More data would help. If we test at the .05 significance level we can show that if we claim superior PIQ for the porpoises there is less than a 5% chance we are wrong.

7-39 $\alpha = .01$ $n = 29$ $\text{df} = 28$ $t_c = \pm 2.76$ $H_0: \mu = 1.9$ $\Sigma x = 61$ $\Sigma x^2 = 153$ $(\Sigma x)^2 = 3{,}721$

$s_{pop} = \sqrt{\dfrac{153 - 3{,}721/29}{28}} = .939$ $s_m = \dfrac{.939}{5.385} = .174$ $m_c = 1.9 \pm (2.76)(.174) =$ $1.9 \pm .48 = 1.42, 2.38$ $m = 61/29 = 2.10$ This value of m is close enough to μ so that we can believe μ is correct. It is not big enough to say μ is wrong with less than a 1% chance of error.

7-45 Expected values

33	60	7	100
66	120	14	200
99	180	21	300 = n

All E's > 5 $\text{df} = 2 \times 1 = 2$ $X_c^2 = 9.21$ $X^2 = 70^2/33 + 30^2/60 + 0/7 + 29^2/66 + 150^2/120 + 21^2/14 - 300 = 395.22 - 300 = 95.22$ Since $95.22 > 9.21$ She can claim that there is a relationship between a student's knowledge of the meaning of impeachment and their attitude. More of those who misunderstand it oppose it than the hypothesis of no relationship leads us to expect. That is, a student who does not know what impeachment means is more likely to be against impeachment than is a person who knows what it means.

7-47 $p = 43/190 = .23$ $s_{dp} = \sqrt{\dfrac{(.23)(.77)}{90} + \dfrac{(.23)(.77)}{100}} = .061$ $p_1 = 20/90 = .22$

$p_2 = 21/100 = .21$ $dp = .03$ $dp_c = 0 \pm 2.58(.061) = \pm .16$ Fail to reject H_0. No evidence that clothing of interviewer was significant.

CHAPTER 8

8-1 **(a)** Let population 1 be the Martians' scores. $n_1 = 6$ $m_1 = 54/6 = 9$ $s_1 = \sqrt{\dfrac{1{,}270 - 2{,}916/6}{5}} = 12.52$ **(b)** $n_2 = 5$ $m_2 = 55/5 = 11$ $s_2 = \sqrt{\dfrac{3{,}925 - 3{,}025/5}{4}} = 28.81$

(c) $s_{dm} = \sqrt{155.75/6 + 830.02/5} = \sqrt{26.125 + 166.004} = 13.86$ **(d)** $\alpha = .01$ $\text{df} = n_1 + n_2 - 2 = 9$ $t_c = \pm 3.25$ $dm_c = 0 \pm (3.25)(13.86) = \pm 45.045$ This sample difference, -2, is not really large enough to show that the mean Martian score is different from the mean Venusian score.

8-3 $\alpha = .01$ $n_1 = 10$ $n_2 = 10$ $\text{df} = 18 \Rightarrow t_c = \pm 2.88$ $m_1 = 13$ $m_2 = 15$ $s_1 = 2$ $s_2 = 3$ $s_{dm} = \sqrt{4/10 + 9/10} = 1.14$ $dm_c = 0 \pm (2.88)(1.14) = \pm 3.28$ $dm = 13 - 15 = -2.$ This difference is not large enough at $\alpha = .01$ to prove that the diets differ.

8-5 Population 1: Male mice who receive male skin Population 2: Female mice who receive male skin $\alpha = .01$ $m_1 = 20$ $m_2 = 14.5$ $s_1 = 1.4$ $s_2 = 1.3$ $n_1 = 20$ $n_2 = 20$ $s_{dm} = \sqrt{1.96/20 + 1.69/20} = .427$ $\text{df} = 38$ $t_c = \pm 2.70$ (closest entry to df = 38)

$dm_c = 0 \pm (2.70)(.427) = 1.15$ $dm = 20 - 14.5 = 5.5$ This is a significant difference. There is very strong evidence that the male-male grafts last longer on average than the male-female grafts. The probability is less than 1 percent that this conclusion is wrong—of course this assumes that both sets of mice were chosen randomly, and that both were equally healthy.

8-7 $n_1 = 8$ $n_2 = 8$ $\text{df} = 14$ $\alpha = .01$ $t_c = \pm 2.98$ $m_1 = 8.5$ $m_2 = 28.5$ $s_1 = 7$ $s_2 = 12$ $s_{dm} = \sqrt{49/8 + 144/8} = 4.91$ $dm_c = 0 \pm (2.98)(4.91) = \pm 14.63$ $dm = 8.5 - 28.5 = -20$ This difference is well beyond the critical values.

8-9 $n_1 = 50 > 30$ $n_2 = 80 > 30$ $\alpha = .01$ $z_c = \pm 2.58$ $m_1 = 630$ $m_2 = 576$ $s_1 = 8$ $s_2 = 5$ $s_{dm} = \sqrt{64/50 + 25/80} = 1.26$ $dm_c = 0 \pm (2.58)(1.26) = \pm 3.25$ This difference is larger than the critical differences. This indicates that probably the vitamin E is causing increased growth.

8-11 $n_1 = 8$ $n_2 = 12$ $\text{df} = 18$ $\alpha = .05$ $t_c = \pm 2.20$ $m_1 = 920/8 = 115$ $m_2 = 1{,}350/12 = 112.5$

$$s_1 = \sqrt{\frac{106{,}400 - 846{,}400/8}{7}} = 9.26 \quad s_2 = \sqrt{\frac{153{,}100 - 1{,}822{,}500/12}{11}} = 10.55$$

$s_{dm} = \sqrt{85.75/8 + 111.30/12} = \sqrt{10.72 + 9.275} = 4.47$ $dm_c = 0 \pm (2.20)(4.47) = \pm 9.834$ $dm = 115 - 112.5 = 2.5$ This is not big enough to show that one class is faster than the other.

8-13 $n_1 = 50$ $n_2 = 50$ $\alpha = .05$ $z_c = \pm 1.96$ $\sigma_1 = 200$ $\sigma_2 = 200$ $\sigma_{dm} = \sqrt{200^2/50 + 200^2/50} = \sqrt{800 + 800} = \sqrt{1{,}600} = 40$ $dm_c = 0 \pm 1.96(40) = \pm 78.4$ The difference between the sample means for the two brands would have to be more than 78.4 hours for us to have a statistically significant result.

8-19 Expected values

20	30	50	100
20	30	50	100
40	60	100	200

E's > 5 $\text{df} = 2 \times 1 = 2$ $\alpha = .05$ $X_c^2 = 5.99$ $X^2 = 18^2/20 + 36^2/30 + 46^2/50 + 22^2/20 + 24^2/30 + 54^2/50 - 200 = 203.44 - 200 = 3.44$ Since $3.44 \ngtr 5.99$ we fail to show that knowledge of who proposed the change affects one's attitude toward the change.

8-21 $p = (17 + 35)/(40 + 67) = 52/107 = .49$ $s_{dp} = \sqrt{\frac{(.49)(.51)}{40} + \frac{(.49)(.51)}{67}} =$ $\sqrt{.0062475 + .003730} = \sqrt{.00998} = .0999$ $dp_c = 0 + (.0999)(\pm 2.58) = \pm .258$ $dp = 17/40 - 35/67 = .425 - .522 = -.097$ The sample difference is too small to be offered as proof of a difference between the beliefs of the males and females.

8-23 $H_0: \mu_{\text{pop}} = 300$ $s_m = 125/\sqrt{60} = 125/7.7459 = 16.14$ $m_c = 300 \pm (1.96)(16.14) = 300 \pm 31.6$ $m_c = 268.4$ and 331.6 $m = 326$ This mean is close enough to the theoretical average so that there is no reason to doubt the claim presented in the null hypothesis.

CHAPTER 9

9-1 **(a)** Binomial one sample estimation (B) **(b)** binomial one sample hypothesis test (A) **(c)** sample mean, two sample, hypothesis test (G) **(d)** sample mean, one sample, hypothesis test (E) **(e)** sample mean, two sample, estimation (H) **(f)** other (I) **(g)** binomial, two sample, hypothesis test (C) **(h)** sample mean, two sample, estimation (H) **(i)** binomial, two sample, estimation (D)

9-3 **(a)** $s_m = 13.2/\sqrt{36} = 13.2/6 = 2.2$, $36 > 30$ $151.8 \pm (1.96)(2.2)$ 151.8 ± 4.31 147.49, 156.11 **(b)** $\Sigma X = 5{,}346.8164$ $m = 148.52267$ $(\Sigma X)^2 = 28{,}588{,}445$

$\Sigma X^2 = 798{,}061.37 \quad s_{pop} = \sqrt{\dfrac{798{,}061.37 - 28{,}588{,}445/36}{35}} = 10.61 \quad s_m = 10.61/6 = 1.77$

$36 > 30$ 95% confidence interval: $148.52 \pm (1.96)(1.77)$ 148.52 ± 3.47 145.05, 151.99 (c) $n = 16$ $\text{df} = 15 \Rightarrow t_c = \pm 2.13$ $s_m = 17.7/4 = 4.425$ $149.9 \pm (2.13)(4.425)$ 149.9 ± 9.43 140.47, 159.33 (d) $\Sigma X = 2{,}480.768$ $(\Sigma X)^2 = 6{,}154{,}209.8$ $m = 155.05$

$\Sigma X^2 = 387{,}542.43 \quad s_{pop} = \sqrt{\dfrac{387{,}542.43 - 6{,}154{,}209.8/16}{15}} = 13.91 \quad s_m = 13.91/4 = 3.48$

95% confidence interval: $155.05 \pm (2.13)(3.48) = 147.64, 162.46$ (e) yes

9-5 $n = 100$ $p = 80/100 = .8$ $s_p = \sqrt{(.8)(.2)/100} = .04$ $80, 20 > 5$ 95% confidence interval: $.8 \pm (1.96)(.04) = .80 \pm .078 = .722, .878$ This interval contains 75% (.75), so we cannot say with 95% confidence that he will win *more than* 75% of his games next year (assuming similar conditions).

9-7 (a) $p = 170/345 = .493$ (b) $s_p = \sqrt{(.493)(.507)/345} = .027$ $170, 175 > 5$, 99% confidence interval: $.493 \pm (2.58)(.027) = .493 \pm .070 = .423, .563$

9-9 (a) $p_1 = 43/50 = .86$ $p_2 = 44/50 = .88$ $dp = -.02$ Two percent more men than women favor the Centers. (b) $s_{dp} = \sqrt{\dfrac{(.86)(.14)}{50} + \dfrac{(.88)(.12)}{50}} = \sqrt{.0024 + .0021} = .067$

$43.7, 44.6 > 5$ 99% confidence interval: $-.02 \pm (2.58)(.067) = -.02 \pm .173 = -.193, .15$

9-11 $p_1 = 52/72 = .722$ $p_2 = 60/81 = .741$ $dp = -.019$ $s_{dp} = \sqrt{\dfrac{(.722)(.278)}{72} + \dfrac{(.741)(.259)}{81}} =$ $\sqrt{.0028 + .0024} = .072$ $52, 20, 60, 21 > 5$ 95% confidence interval: $-.019 \pm (1.96)(.072) = -.019 \pm .141 = -.16, .122$

9-13 (a) $m = 63.8$ (b) $s_{pop} = 2.2$ $s_m = 2.2/\sqrt{40} = .348$ $40 > 30$ 95% confidence interval: $63.8 \pm (1.96)(.348) = 63.8 \pm .682 = 63.12, 64.48$

9-15 $m = \$13{,}000$ $s_{pop} = 1{,}400$ $s_m = 1{,}400/\sqrt{49} = \$200$ $49 > 30$ 99% confidence interval: $13{,}000 \pm (2.58)(200) = 13{,}000 \pm 516 = \$12{,}484, \$13{,}516$

9-17 $m = 4.3$ $s_{pop} = 6$ $s_m = 6/\sqrt{23} = 1.25$ $\text{df} = 22$ $t_c = \pm 2.82$ 99% confidence interval: $4.3 \pm (2.82)(1.25) = 4.3 \pm 3.525 = .775, 7.825$

9-19

X	4	2	1	4	3	2	4	5	3	3	3	4	0	1
X^2	16	4	1	16	9	4	16	25	9	9	9	16	0	1

X	1	2	2	1	4	5	3	2	0	1	2
X^2	1	4	4	1	16	25	9	4	0	1	4

$\Sigma X = 62 \quad \Sigma X^2 = 204 \quad s_{pop} = \sqrt{\dfrac{204 - 3{,}844/25}{24}} = 1.45 \quad s_m = 1.45/5 = .29$

$m = 62/25 = 2.48$ $\text{df} = 24$ $t_c = \pm 2.80$ 99% confidence interval: $2.48 \pm (2.80)(.29) = 2.48 \pm .812 = 1.67, 3.29$

9-21 $m_1 = 2.6$ $m_2 = 2.3$ $s_1 = 1.8$ $s_2 = 1.6$ $n_1 = 31$ $n_2 = 33$ $dm = .3$ $s_{dm} = \sqrt{3.24/31 + 2.56/33} = \sqrt{.1045 + .0776} = .427$, 31 and 33 > 30 99% confidence interval: $.3 \pm (2.58)(.427) = .3 \pm 1.10 = -.8, 1.4$ This interval contains 0. Therefore we cannot say that we have 99% confidence that the men want more children than the women do.

9-23 $m_1 = 80.1$ $m_2 = 81.2$ $s_1 = 6.3$ $s_2 = 7.0$ $n_1 = 40$ $n_2 = 20$ $dm = -1.1$ $\text{df} = 58$ $t_c = \pm 2.00$ $s_{dm} = \sqrt{39.69/40 + 49/20} = \sqrt{.9923 + 2.45} = 1.86$ 95% confidence interval: $-1.1 \pm (2.00)(1.86) = -1.1 \pm 3.72 = -4.82, 2.62$

9-25 Population 1: Boys in hospitals $m_1 = 9.7$ $m_2 = 9.9$ $s_1 = .2$ $s_2 = .1$ $n_1 = 20$ $n_2 = 20$ $dm = -.2$ $\text{df} = 38$ $t_c = \pm 2.02$ $s_{dm} = \sqrt{.04/20 + .01/20} = .05$ 95% confidence interval: $-.2 \pm (2.02)(.05) = -.2 \pm .101 = -.301, -.099$ Evidently, since 0 is not in this interval, girls in hospitals have, on the average, more fingers than boys do.

9-27 $m_1 = 73$ $m_2 = 63$ $s_1 = 10$ $s_2 = 10$ $n_1 = 50$ $n_2 = 20$ $dm = 10$ $df = 68$ $t_c = \pm 2.66$ $s_{dm} = \sqrt{100/50 + 100/50} = 2$ 99% confidence interval: $10 \pm (2.66)(2) = 10 \pm 5.32 =$ 4.68, 15.32 This indicates that the history majors are more aware of current events than nonhistory majors—assuming of course that we have two *random* samples, not, for example, just history majors who specialize in contemporary civilization.

9-29 (a)

Toss No.	1	2	3	4	5	6	7	8	9	10	11	12	13
Percent	70	60	53	50	52	50	53	51	52	52	54	53	54

Toss No.	14	15
Percent	56	53

(b) $n = 15 \times 10 = 150$ $s_p = \sqrt{(.53)(.47)/150} = .0408$ outcome: 80 up, 70 down are both > 5; therefore, nd_p, $z = 1.96$ $p \pm zs = .53 \pm 1.96(.0408) = .53 \pm .08 = .45$ and .61; therefore, $45\% < \pi < 61\%$

9-31 (b) 434.5 and 115.5 > 5 $.79 \pm 1.96\sqrt{(.79)(.21)/550} = .79 \pm 1.96(.0174) = .79 \pm .03 = .76$ and .82 (c) $.79 \pm 2.58(.0174) = .79 \pm .044 = .746$ and .834 (d) We are 95% sure $\pi > .75$ (that is, $.76 < \pi < .82$). However we are *not* 99% sure of this.

9-39 *Solution taken from a student project* Population: Male shoppers of Automotive Discount Center over a 3-day span, Thursday through Saturday (peak shopping days) Hypothesis: Let $\pi = P$(a person polled uses Brand B spark plugs) $H_0: \pi = .60$ $H_a: \pi \neq .60$ Sample: I polled the first 293 who came into the store. It took me 293 men to get 200 to be able to fit my requirements. First of all, I left out women because they usually do not do their own tune-ups and I do not think the exceptions would have made any difference anyway. Secondly, they all had to own a G.M. car; they would have to do their own spark plug changes. Out of 293 men 61 did not own a G.M. car and 32 did not change their own spark plugs. All 293 answered my questions, "Excuse me sir, do you own a car made by G.M. and if so, do you change your own spark plugs?" This question can be answered by a simple yes or no. Out of the 200, 109 used Brand B and 91 used Brand A. My store only carries Brands A and B. $n = 200$ $p = 109/200 = .54$ Significance level: Let $\alpha = .05$, $n\pi = 200(.60) = 120$ $1 = \pi = .40$, $n(1 = \pi) =$ $200(.40) = 80$ $\left.\begin{matrix}80\\120\end{matrix}\right\} > 5$ So, we can use the normal distribution to approximate the binomial distribution, and the critical z scores will be ± 1.96 mean: $\mu_p = .60$ standard deviation: $\sigma_p = \sqrt{\pi(1-\pi)/n} = \sqrt{(.60)(.40)/200} = \sqrt{.24/200} = \sqrt{.0012} = .0346$ $nd_p(\mu = .60, \sigma = .0346)$ Critical values: $p_c = \mu + z_c\sigma = .60 + (\pm 1.96)(.0346) =$ $.60 + (\pm .07) = .53$ and .67 Decision rule: I will reject the null hypothesis. If my sample value of p is less than .53 or greater than .67. I have learned from my computations that p will almost always be between 53% and 67% if π is really 60%. Outcome: From my previous sample the outcome was $p = .54$ Decision: My experimental result was $p = .54$. Therefore I fail to reject the null hypothesis. I have not proved at the .05 significance level that 60% of G.M. cars privately purchased was too high a percentage for switching to Brand B. My computations tell me that it is not unusual to get a sample of $p = .54$ when the true value of the parameter is $\pi = .60$. Comments: It is true most individuals get their car tuned up at either a gas station or from their original dealers but I do not take this fact into consideration because there are at least 6 types of spark plugs. The original advertisement made by Brand B stated that the reason for the change was for better performance, durability, gas mileage, etc. Service stations and dealers use whatever type of plug they can get the best deal and/or price on. So, its possible to poll six different service stations or dealers and get six different types of plugs used because of company policy, package class, etc. The Saturday mechanic who patronizes my store can only pick Brand A or B (the price is the same for both) so their preference is determined by past performance, durability, etc. Since price does not put forth a factor, they choose their plugs with the same reasons Brand B says they do. This is why I feel my poll is relatively accurate.

9-41 Population: (parttime, night employees at Gron's) $n = 13$; Therefore use t distribution

$df = 12$, so $t_c = 2.18$ $H_0: \mu_{pop} = \$2.30$ $H_a: \mu_{pop} \neq \$2.30$ $s_{pop} = \sqrt{\frac{\Sigma X^2 - (\Sigma X)^2/n}{n-1}} =$

$\sqrt{\frac{62.0575 - (28.25)^2/13}{12}} = \sqrt{\frac{62.0575 - 61.3894}{12}} = \sqrt{.6681/12} = \sqrt{.05568} = .236$

X	2.10	2.10	2.00	2.00	2.00	2.00	2.00	2.15	2.05	2.20	2.75	2.40	2.50
X²	4.41	4.41	4.00	4.00	4.00	4.00	4.00	4.6225	4.2025	4.84	7.5625	5.76	6.25

$\Sigma X = 28.25$ $\Sigma X^2 = 62.0575$ $s_m = 236/\sqrt{13} = .236/3.606 = .065$ $13 \not> 30$; therefore use t distribution td_m ($\mu = 2.30$, $s = .065$) $m_c = \mu_m + s_m t_c = 2.30 + (\pm 2.18)(.065) = 2.30 + (\pm.1417) = 2.16$ and 2.44 Decision rule: I will reject H_0 if my outcome is <2.16 or >2.44. Outcome: $m = 2.17$ Conclusion: I cannot reject the manager's claim on the basis of this evidence. However, the results are so close that it would be valuable to collect a larger sample and repeat the calculations.

CHAPTER 10

10-1 (a) Looks like some weak positive correlation. (b) $\Sigma XY = 370$ (c) $\Sigma X = 51$ (d) $\Sigma Y = 35$ (e) $\Sigma X^2 = 639$ (f) $\Sigma Y^2 = 255$ (g) $n = 5$

(h) $$\frac{(5)(370) - (51)(35)}{\sqrt{(5)(639) - (51)^2}\sqrt{(5)(255) - (35)^2}} = \frac{1,850 - 1,785}{(24.37)(7.07)} = .377$$

	X	Y	X²	Y²	XY
	14	6	196	36	84
	3	7	9	49	21
	8	5	64	25	40
	9	8	81	64	72
	17	9	289	81	153
Total	51	35	639	255	370

(i) $H_0: \rho = 0$ $H_a: \rho \neq 0$ $r_c = \pm.88$ Since $-.88 < .377 < .88$ we fail to reject H_0. There may be no correlation between the finger spans of Venusians' left hands and their right hands.

10-3

X	Y	XY	X²	Y²
13	1	13	169	1
17	1	17	289	1
19	3	57	361	9
19	1	19	361	1
20	4	80	400	16
25	5	125	625	25
39	1	39	1,521	1

(a) Not much correlation (b) $\Sigma XY = 350$ (c) $\Sigma X = 152$ (d) $\Sigma Y = 16$ (e) $\Sigma X^2 = 3{,}726$ (f) $\Sigma Y^2 = 54$ (g) $n = 7$ (h) $r = \dfrac{(7)(350) - (152)(16)}{\sqrt{(7)(3{,}726) - (152)^2}\ \sqrt{(7)(54) - (16)^2}} =$

$(2{,}450 - 2{,}432)/(54.57)(11.05) = .030$

10-5 (a) Let X = number of children. Let Y = number of pets $n = 15$

See Table A10-5. $\Sigma X = 32$ $\Sigma Y = 13$ $\Sigma XY = 38$ $\Sigma X^2 = 124$ $\Sigma Y^2 = 21$

$$r = \frac{(15)(38) - (32)(13)}{\sqrt{(15)(124) - (32)^2}\ \sqrt{(15)(21) - (13)^2}} = \frac{570 - 416}{(28.91)(12.08)} = .44$$

(b) $\alpha = .01$ $r_c = \pm.64$ r is not large enough for us to claim with 99% confidence that there is actually correlation in the population.

TABLE A10-5

X	Y	XY	X^2	Y^2
0	0	0	0	0
3	1	3	9	1
1	0	0	1	0
5	0	0	25	0
0	1	0	0	1
2	1	2	4	1
6	3	18	36	9
1	1	1	1	1
5	1	5	25	1
2	0	0	4	0
3	2	6	9	4
1	0	0	1	0
0	1	0	0	1
3	1	3	9	1
0	1	0	0	1

10-7

X	Y	XY	X^2	Y^2
40	28	1,120	1,600	784
36	30	1,080	1,296	900
24	22	528	576	484
42	28	1,176	1,764	784
30	24	720	900	576
35	26	910	1,225	676
21	18	378	441	324
34	29	986	1,156	841
34	22	748	1,156	484
32	30	960	1,024	900

$n = 10$ $\Sigma X = 328$

$\Sigma Y = 257$ $\Sigma X^2 = 11{,}138$

$\Sigma XY = 68{,}606$ $\Sigma Y^2 = 6{,}753$

$$r = \frac{(10)(8{,}606) - (328)(257)}{\sqrt{(10)(11{,}138) - (328)^2}\ \sqrt{(10)(6753) - (257)^2}} = \frac{86{,}060 - 84{,}296}{(61.61)(38.48)} = .744 \quad \alpha = .01$$

$r_c = \pm.76$ r is not large enough for us to say (with $\alpha = .01$) that we have found correlation in the population. Since r is close to r_c, it would be useful to collect more data to see if the trend stays the same.

10-9 See Table A10-9. $r = \dfrac{14(410{,}387) - 1{,}616(2{,}975)}{\sqrt{(14)(293{,}660) - 1616^2}\ \sqrt{(14)(738{,}821) - 2{,}975^2}} = .63$

TABLE A10-9

X	Y	XY	X^2	Y^2
66	127	8,382	4,356	16,129
27	147	3,969	729	21,609
204	204	41,616	41,616	41,616
246	453	111,438	60,516	205,209
206	359	73,954	42,436	128,881
291	230	66,930	84,681	52,900
67	133	8,911	4,489	17,689
31	234	7,254	961	54,756
28	150	4,200	784	22,500
80	195	15,600	6,400	38,025
185	191	35,335	34,225	36,481
85	157	13,345	7,225	24,649
61	184	11,224	3,721	33,856
39	211	8,229	1,521	44,521

$r_c = \pm.53$ Reject $\rho = 0$. There is some positive correlation.

10-11 $n = 30$ $\alpha = .01 \Rightarrow r_c = \pm.46$ We have $r = .83$. This is good evidence for claiming that ages and grades are positively correlated. The older students tend to get the higher grades.

10-13 See Table A10-13. $n = 10$ $\Sigma X = 110$ $\Sigma Y = -72$ $\Sigma XY = -4{,}188$ $\Sigma X^2 = 1{,}540$

$$\Sigma Y^2 = 35{,}728 \quad r = \frac{10(-4{,}188) - (110)(-72)}{\sqrt{10(1{,}540) - (110)^2}\sqrt{10(35{,}728) - (-72)^2}} = \frac{4{,}188 + 7{,}920}{(57.45)(593.38)} = -10.29$$

$$m_X = 110/10 = 11 \quad m_Y = -72/10 = -7.2 \quad b = \frac{(10)(-4188) - (110)(-72)}{10(1540) - (110)^2} =$$

$$\frac{-41{,}880 + 7{,}920}{3{,}300} = -10.29$$

TABLE A10-13

X	Y	X^2	Y^2	XY
2	80	4	6,400	160
4	70	16	4,900	280
6	42	36	1,764	252
8	32	64	1,024	256
10	−06	100	36	−60
12	−76	144	256	−192
14	−34	196	1,156	−476
16	−56	256	3,136	−896
18	−84	324	7,056	−1,512
20	−100	400	10,000	−2,000

10-15

X	Y	X^2	Y^2	XY
4	2	16	4	8
2.5	1	6.25	1	2.5
4.5	2	20.25	4	9.0
3	1.5	9	2.25	4.5
4	2	16	4	8

$n = 5$ $\Sigma X^2 = 67.5$ $m_X = 18/5 = 3.6$
$\Sigma X = 18$ $\Sigma Y^2 = 15.25$ $m_Y = 8.5/5 = 1.7$
$\Sigma Y = 8.5$ $\Sigma XY = 32$

(a) $b = [(5)(32) - (18)(8.5)]/[5(67.5) - (18)^2] = (160 - 153)/13.5 = .519$
(b) $Y = b(X - m_X) + m_Y = .519(X - 3.6) + 1.7$ **(c)** $Y = .519(3.5 - 3.6) + 1.7 = 1.65$
(d) $r = 7/(3.67)(2) = .954$ $r_c = \pm.88$ at $\alpha = .05$ **(e)** This is strong correlation, so the prediction will probably be close to correct.

10-17

X	Y	X^2	Y^2	XY
14	6	196	36	84
3	7	9	49	21
8	5	64	25	40
9	8	81	64	72
17	9	289	81	153

$n = 5$ $\Sigma Y^2 = 255$
$\Sigma X = 51$ $\Sigma XY = 370$
$\Sigma Y = 35$ $m_X = 10.2$
$\Sigma X^2 = 639$ $m_Y = 7$

(a) $b = \dfrac{5(370) - 51(35)}{5(639) - (51)^2} = \dfrac{1{,}850 - 1{,}785}{594} = .109$ **(b)** $\tilde{Y} = .109(X - 10.2) + 7$

(c) $\tilde{Y} = .109(10 - 10.2) + 7 = 6.98$

10-19 $\tilde{Y} = \dfrac{18}{2{,}978}\left(19 - \dfrac{152}{7}\right) + \dfrac{16}{7} = 2.3$

10-21 $\tilde{Y} = (154/836)(4 - 32/15) + 13/15 = 1.2$

10-23 $\tilde{Y} = (1{,}764/3{,}796)(34 - 32.8) + 25.7 = 26.3$

10-33 H_0: Sex and smoking is related. H_a: Sex and smoking is not related.
$\alpha = .05$ 2×4 $\text{df} = 1 \times 3 = 3$ $X_c^2 = 7.82$

O	E	O^2	O^2/E
6	7.02	36	5.13
16	14.58	256	17.56
26	24.3	676	27.82
6	8.1	36	4.44
7	5.98	49	8.19
11	12.42	121	9.74
19	20.7	361	17.44
9	6.9	81	11.74
Total			102.06
			−100
Difference			$2.06 = X^2$

Decision Rule: I will reject the null hypothesis if my outcome is greater than 7.82. Outcome: $X^2 = 2.06$ Conclusion: I fail to reject the null hypothesis. There may be no relationship between sex and smoking. Males and females are equally likely to fit any of the categories.

10-35 $p = (909 + 61)/(1{,}057 + 100) = 970/1{,}157 = .84$ $s_{dp} = \sqrt{\dfrac{(.84)(.16)}{1{,}057} + \dfrac{(.84)(.16)}{100}} +$
$\sqrt{.000127 + .001344} = \sqrt{.00147} = .03834$ $dp_c = 0 + (\pm 2.58)(.03834) = \pm.099$
$dp = .86 - .61 = .25$ The sample difference is large enough to be considered good evidence that the two populations are different. For some reason fewer of the local men are comfortable in their role.

10-37 $H_0: \mu_{dm} = 0$ $s_{dm} = \sqrt{\dfrac{40{,}000^2}{50} + \dfrac{50{,}000^2}{20}} = \sqrt{32{,}000{,}000 + 125{,}000{,}000} = \sqrt{157{,}000{,}000} =$

12,529.964 = 12,530 df = 49 + 19 = 68 ⇒ $t_c = 2.00$ $dm_c = 0 + (\pm 2.00)(12{,}530) = \pm 25{,}060$ $dm = 186{,}000 - 150{,}000 = 36{,}000$ The sample difference is big enough to indicate that the Venusian ships travel more per year than the Saturnian ones do.

CHAPTER 11

11-1 (a) Mean rank = $(1 + 2 + 3 + \cdots + 12)/12 = 78/12 = 6.5$ (b) $5(6.5) = 32.5$
(c) $5(5 + 7 + 1)/2 = 65/2 = 32.5$ same as part b

11-3 H_0: Number of meals missed is the same for both groups

$n_1 = 7$		$n_2 = 8$	
Couples on First Week		Couples on Second Week	
Number of Meals Missed	Rank	Number of Meals Missed	Rank
5	3	0	1
7	6	3	2
7	6	6	4
8	8.5	7	6
10	11.5	8	8.5
10	11.5	10	11.5
17	15	10	11.5
		13	14
	61.5 = W		

$W_c = 38$ and 74 $38 < 61.5 < 74$ Fail to reject H_0

11-5

Ranks for Country X	Ranks for Country Z
14	7
1.5	5
8	3
11.5	9.5
16	17.5
11.5	15
6	9.5
13	17.5
4	1.5
W = 85.5	

$\alpha = .05$
$n_{small} = 9$
$n_{large} = 9$
$W_C - 63{,}108$

Since W is neither exceptionally small nor large we cannot conclude that either country does a superior job.

11-7

Days to Recover Treatment A	Days to Recover Treatment B	Ranks A	Ranks B
5	8	1.5	7
6	8	3.5	7
7	9	5	10
5	10	1.5	12
8	9	7	10
6	9	3.5	10
		W = 22	

$\alpha = .05$ $n_A = 6$ $n_B = 6 \Rightarrow W_c = 26, 52$ Since W is smaller than the critical value 23, we have strong evidence that Treatment A is quicker than Treatment B. If A and B were equally quick, such results as we got would be extremely rare, occurring less than 5% of the time.

11-9 (a) $n_1 = n_2 = 60$ $\mu_W = 60(121)/2 = 3{,}630$ $\sigma_W = \sqrt{\dfrac{60(60)(121)}{12}} = 190.5$

$W_c = 3{,}630 \pm 2.58(190.5) = 3{,}138;\ 4{,}122$ This does not indicate that there is a difference at the .01 significance level.

11-11 See Table A11-11. $S = 414$ $r_S = 1 - 6(414)/20(399) = .689$ $\alpha = .05$ $r_c = \pm 1.96/\sqrt{19} = \pm .450$ r_S is large enough to support the idea that "the closer you sit, the higher your grade tends to be."

TABLE A11-11

Student	Row Rank	Grade Rank	Difference Squared
1	14	15.5	2.25
2	14	9.5	20.25
3	2.5	3.5	1.00
4	2.5	9.5	49.00
5	14	3.5	110.25
6	18.5	15.5	9.00
7	6.5	3.5	9.00
8	6.5	9.5	9.00
9	2.5	3.5	1.00
10	18.5	19.5	1.00
11	6.5	3.5	9.00
12	18.5	15.5	9.00
13	10	15.5	30.25
14	18.5	9.5	81.00
15	2.5	3.5	1.00
16	6.5	9.5	9.00
17	14	19.5	30.25
18	14	15.5	2.25
19	10	9.5	.25
20	10	15.5	30.25

11-13 See Table A11-13. $S = 1105.50$ $r_S = 1 - 6(1{,}105.50)/20(399) = .169 = .05 \Rightarrow r_c = \pm 1.96/\sqrt{19} = \pm .45$ We see that r_S is not significantly different from 0. We have no strong evidence from this sample that there is a correlation among athletes between calf muscle elasticity and high jump performance.

TABLE A11-13

Athlete	Difference of Ranks, Squared	Athlete	Difference of Ranks, Squared
1	36	11	100
2	0	12	4
3	81	13	1
4	36	14	9
5	144	15	156.25
6	36	16	110.25
7	1	17	169
8	144	18	4
9	36	19	9
10	25	20	4

11-15

Package	Masculinity Rank	Popularity Rank	Difference Squared
1	1	2	1
2	2	1	1
3	3	4	1
4	4	7	9
5	5	3	4
6	6	6	0
7	7	8	1
8	8	5	9

$S = 26$ $r_S = 1 - 6(26)/8(63) = .25$ $\alpha = .05 \Rightarrow r_c = -.69$ and $.71$ The correlation in the sample is not strong enough to indicate correlation in the population. Notice though that by converting the "number of samples taken" to ranks, we have thrown away a lot of information. For example, it was clear from the original data that packages 1 and 2 were a *lot* more popular than most, but this information is lost when we convert to rank scores. It may be useful to try to establish a "masculinity" measure which was stronger than just a rank measure—say, score each package for masculinity on a scale from 1 to 100. Then we could perform an ordinary Pearson correlation test using all the information.

11-17

Colony	1	2	3	4	5	6	7	8
Difference of Ranks, Squared	.25	.25	1	1	0	1	0	1

$S = 4.5$ $r_s = 1 - 6(4.5)/8(63) = .946$ $\alpha = .05 \Rightarrow r_c = -.69$ and $.71$ This indicates a very strong correlation between crowding and incidents of violence in laboratory rat colonies.

11-19 (a) See Table A11-19a. $\Sigma X = 305$ $\Sigma Y = 2{,}159$ $\Sigma X^2 = 4{,}759$ $\Sigma Y^2 = 158{,}943$

TABLE A11-19a

X = Number of Hours Studied	Y = Grade on Exam	X^2	Y^2	XY
21	95	441	9,025	1,995
18	72	324	5,184	1,296
1	61	1	3,721	61
9	73	81	5,329	657
13	68	169	4,624	884
1	52	1	2,704	52
17	60	289	3,600	1,020
2	86	4	7,396	172
19	67	361	4,489	1,273
1	64	1	4,096	64
6	73	36	5,329	438
19	76	361	5,776	1,444
14	53	196	2,809	742
4	69	16	4,761	276
0	90	0	8,100	0
20	70	400	4,900	1,400
18	86	324	7,396	1,548
2	75	4	5,625	150
12	75	144	5,625	900
20	59	400	3,481	1,180
15	66	225	4,356	990
11	67	121	4,489	737

TABLE A11-19a (*Continued*)

X = Number of Hours Studied	Y = Grade on Exam	X²	Y²	XY
21	84	441	7,056	1,764
14	78	196	6,084	1,092
13	81	169	6,561	1,053
1	75	1	5,625	75
3	86	9	7,396	258
6	81	36	6,561	486
2	59	4	3,481	118
2	58	4	3,364	116

$\Sigma XY = 22{,}241 \quad n = 30 \quad r = \dfrac{30(22{,}241) - (305)(2{,}159)}{\sqrt{30(4{,}759) - (305)^2}\ \sqrt{30(158{,}943) - (2{,}159)^2}} =$ $\dfrac{667{,}230 - 658{,}495}{(223.04)(327.12)} = .12 \quad r_c = .36$ No significant correlation between study time and grade on exam. (**b**) Let 1 be the rank for the *most* hours studied. Let 1 be the rank for the *highest* grade. See Table A11.19b. $S = 3910.5 \quad r_S = 1 - 6(3{,}910.5)/30(899) = .13$ $r_c = \pm 1.96/\sqrt{29} = \pm .36$ Since r_S is between the two values of r_c, we have no solid evidence of correlation between grades on the exam and time spent studying for it during the last week.

TABLE A11-19b

Student	Hours Rank	Grade Rank	Difference Squared
1	1.5	1	.25
2	7.5	16	72.25
3	27.5	24	12.25
4	17	14.5	6.25
5	13.5	19	30.25
6	27.5	30	6.25
7	9	25	256
8	23.5	4	380.25
9	5.5	20.5	225
10	27.5	23	20.25
11	18.5	14.5	16
12	5.5	10	20.25
13	11.5	29	306.25
14	20	18	4
15	30	2	784
16	3.5	17	182.25
17	7.5	4	12.25
18	23.5	12	132.25
19	15	12	9
20	3.5	26.5	529
21	10	22	144
22	16	20.5	20.25
23	1.5	6	20.25
24	11.5	9	6.25
25	13.5	7.5	36
26	27.5	12	240.25
27	21	4	289
28	18.5	7.5	121
29	23.5	26.5	9
30	23.5	28	20.25

11-21

Ranks for Horses Who Had Drug	Ranks for Horses Who Had No Drug
2	10
1	7
4	9
6	8
3	5

$W = 16$ $\alpha = .05 \Rightarrow W_c = 17, 38$ Since W is less than the smaller critical value, we see that W is too low to have occurred just by chance. It seems evident that the drug does have an effect on the performance of the horses.

11-27 $H_0 : \pi = P$(a student will go on to a four year school) $= .60$ $n = 85$ $n\pi = 51 > 5$ $n(1 - \pi) = 34 > 5$ Therefore, nd_p. $\mu = .6$ $\sigma = \sqrt{(.6)(.4)/85} = .053$ $p_c = .6 \pm 1.96(.053) = .50$ and $.70$ $p = 36/85 = .42$. Reject H_0. 60% is too high

11-29 $H_0 : \mu_{dm} = \mu_1 - \mu_2 = 0$ $90, 50 > 30$ Therefore, nd_{dm}. $s_{dm} = \sqrt{11.3^2/90 + 13.2^2/50} = 2.21$ $dm = 0 \pm 1.96(2.21) = \pm 4.3$ $dm = 16.8 - 20.6 = -3.8$ Fail to reject H_0. The two sexes may smoke the same number of cigarettes on average.

11-31 (a) $p_1 = 80/100 = .80$ $p_2 = 75/100 = .75$ $p_1 - p_2 = .80 - .75 = .05$ (b) $s_{dp} = \sqrt{\frac{(.8)(.2)}{100} + \frac{(.75)(.25)}{100}} = .0589$ $80, 20, 75, 25 > 5$ $.05 \pm 2.58(.0589) = -.10$ and $+.20$ $-.10 < \pi_1 - \pi_2 < .20$

APPENDIX E

E-1 (a) $p_c = .25 \pm 1.96(.03) = .191, .309$

(b) $$z_{.309} = \frac{.309 - .40}{.035} = -2.6$$

$$z_{.191} = \frac{.191 - .40}{.035} = -6.0$$

$$\beta = .0047 - .0000 = .0047$$

E-3 (a) $m_c = 6500 \pm 1.96(12.6) = 6475, 6525$

(b) $$z_{6525} = \frac{6525 - 6000}{12.6} = 42$$

$$z_{6475} = \frac{6475 - 6000}{12.6} = 37$$

$$\beta = 0 - 0 = 0$$

INDEX

α (alpha), 49
Alternative hypothesis, 50
Averages, 124–153

b, slope of best fitting line, 204
Binomial experiment, 10, 89
Binomial hypothesis test, 96
Binomial (2-sample) test, 107–119

Central limit theorem, 133
Chi square, 10, 40–64
Cicero, 24
Coin toss, 7
Computer simulation, 7
Confidence interval, 176
Contingency table, 57
Correction factor for χ^2, 73
Correlation, 193
Cowper, William, 24
Critical value, 45, 48

David, F. N., 24
Degrees of freedom, 43, 59, 61
Dependence, 61
Descartes, René, 40
Deviation, 93
Difference of means, 157–167
Distribution free test, 209–232
dm, 160
dp, 110

Error:
  Type I, 51
  Type II, 51
Estimation, 173–183
Expected values, 41

Frequency graph, 17
Frequency table, 17

Greek letter, use of, 35

H_a (alternative hypothesis), 50
H_0 (null hypothesis), 50
Hypothesis, 4, 49
Hypothesis test (averages), 129
Hypothesis test (sample means), 158
Hypothesis test (summary), 152

Independence, 56
Inference, 14
Interval estimate, 175

m, 76
Mean, 126
Mean (of normal distribution), 78
Median, 126
Midrange, 129
Mode, 33, 129
μ, 76

nd, 82
Nonparametric test, 210
Normal curve, 77
Null hypothesis, 48, 50

Observed values, 41

p, 94
Parameter, 76
Pearson, Karl, 44
Percentages, 109
π, 35
Point estimate, 174
Population, 4
Prediction, 202
Probability, 24–35

r, 195
Random experiment, 3
Random sample, 12–21
Range, 79
Rank correlation, 226
Rank sum, 214
Rank test, 211

s, 135
Sample, 4
Sample differences, 110

Sample mean, 10, 129, 130
Scattergram, 192
Significance level, 48, 49
Simulation, 7
Small sample, 144
Spearman, 226
Standard deviation, 79
Standard deviation (estimate), 135
Statistic, 50, 76
Statistical hypothesis, 4
Symbols, 76

t curve, 148
td, 148
Test statistic, 41

Variability, 79

Wilcoxon, 215

z score, 81

SYMBOLS

α [alpha]	Probability of Type I error
β [beta]	Probability of Type II error
μ [mu]	Population mean
π [pi]	A probability
ρ [rho]	Correlation coefficient
σ [sigma, lowercase]	Population standard deviation
Σ [sigma, uppercase]	The command to add
H_a [H sub a]	Alternative hypothesis
H_0 [H sub 0]	Null hypothesis
z_c [z sub c]	Critical z score
t_c [t sub c]	Critical t score
X	Raw score
n	Sample size
s_{pop}	Estimate of population standard deviation
m	Estimate of population mean
P(event)	Probability of an event
df	Degrees of freedom
dm	Difference of means
dp	Difference of proportions
r	Estimate of correlation coefficient
nd_X ($\mu = 20$, $\sigma = 4$)	A normal distribution of X's with mean equal to 20 and standard deviation equal to 4
r_S	Estimate of coefficient of correlation for ranks
S	Sum of squares
W	Rank sum of the smaller sample

FORMULAS

Formula	Description
$m = \dfrac{\Sigma X}{n}$	Sample mean
$z = \dfrac{X - \mu}{\sigma}$	z score from raw score
$X = \mu + z\sigma$	Raw score from z score
$s_{\text{pop}} = \sqrt{\dfrac{\Sigma X^2 - \dfrac{(\Sigma X)^2}{n}}{n - 1}}$	Estimate of population standard deviation from raw scores
$\mu_p = \pi$	Mean of the p's (binomial)
$\sigma_p = \sqrt{\dfrac{\pi(1 - \pi)}{n}}$	Standard deviation of the p's
$s_p = \sqrt{\dfrac{pq}{n}}$	Estimate of the standard deviation of the p's
$p = \dfrac{f}{n}$	Estimate of π, the proportion of pop in a particular group
$p = \dfrac{f_1 + f_2}{n_1 + n_2}$	Estimate of π under assumption that $\pi_1 = \pi_2 = \pi$
$s_{dp} = \sqrt{\dfrac{pq}{n_1} + \dfrac{pq}{n_2}}$	Estimate of standard deviation of differences of proportions assuming $\pi_1 = \pi_2 = \pi$
$s_{dp} = \sqrt{\dfrac{p_1q_1}{n_1} + \dfrac{p_2q_2}{n_2}}$	Estimate of standard deviation of differences of proportions not assuming $\pi_1 = \pi_2$
$\sigma_m = \dfrac{\sigma_{\text{pop}}}{\sqrt{n}}$	Standard deviation of sample means
$s_m = \dfrac{s_{\text{pop}}}{\sqrt{n}}$	Estimate of standard deviation of sample means
$s_{dm} = \sqrt{\dfrac{{s_1}^2}{n_1} + \dfrac{{s_2}^2}{n_2}}$	Estimate of standard deviation of differences of means
$p - z_c s_p < \pi < p + z_c s_p$	Confidence interval for π
$m - z_c s_m < \mu < m + z_c s_m$	Confidence interval for μ, $n > 30$
$m - t_c s_m < \mu < m + t_c s_m$	Confidence interval for μ, $n \ngtr 30$
$X^2 = \Sigma \dfrac{O^2}{E} - n$	X^2 statistic
$r = \dfrac{n\Sigma XY - (\Sigma X)(\Sigma Y)}{\sqrt{n\Sigma X^2 - (\Sigma X)^2}\,\sqrt{n\Sigma Y^2 - (\Sigma Y)^2}}$	Estimate of coefficient of correlation
$\mu_W = \dfrac{n_1(n_1 + n_2 + 1)}{2}$	Mean of the rank sums
$\sigma_W = \sqrt{\dfrac{n_1n_2(n_1 + n_2 + 1)}{12}}$	Standard deviation of the rank sums